Human Microbiome

Edited by Natalia V. Beloborodova and Andrey V. Grechko

Published in London, United Kingdom

IntechOpen

Supporting open minds since 2005

Human Microbiome
http://dx.doi.org/10.5772/intechopen.85279
Edited by Natalia V. Beloborodova and Andrey V. Grechko

Part of IntechOpen Book Series: Biochemistry, Volume 20
Book Series Editor: Miroslav Blumenberg

Contributors
Andra-Iulia F. Suceveanu, Marilena Musat, Andrada Dumitru, Claudia Voinea, Felix Voinea, Irinel Parepa, Anca Pantea Stoian, Adrian-Paul Suceveanu, Laura Mazilu, Akhlash P. Singh, Najaf Allahyari Fard, Prasat Kittakoop, Teresa Auguet, Laia Bertran, Jessica Binetti, Jitlada Meephansan, Hok Bing Thio, Atiya Rungjang Rungjang, Natalia V. Beloborodova, Andrey V. Grechko

First published in London, United Kingdom, 2021 by IntechOpen
IntechOpen is the global imprint of INTECHOPEN LIMITED, registered in England and Wales, registration number: 11086078, 5 Princes Gate Court, London, SW7 2QJ, United Kingdom
Printed in Croatia

British Library Cataloguing-in-Publication Data
A catalogue record for this book is available from the British Library

Additional hard and PDF copies can be obtained from orders@intechopen.com

Human Microbiome
Edited by Natalia V. Beloborodova and Andrey V. Grechko
p. cm.
Print ISBN 978-1-78984-848-9
Online ISBN 978-1-78984-849-6
eBook (PDF) ISBN 978-1-78985-261-5
ISSN 2632-0983

For EU product safety concerns:
IN TECH d.o.o., Prolaz Marije Krucifikse Kozulić 3, 51000 Rijeka, Croatia, info@intechopen.com or visit our website at intechopen.com.

IntechOpen Book Series

Biochemistry

Volume 20

Natalia Vladimirovna Beloborodova was educated at the Pirogov Russian National Research Medical University, with a degree in pediatrics in 1980, a Ph.D. in 1987, and a specialization in Clinical Microbiology from First Moscow State Medical University in 2004. She has been a Professor since 1996. Currently, she is the Head of the Laboratory of Metabolism, a division of the Federal Research and Clinical Center of Intensive Care Medicine and Rehabilitology, Moscow, Russian Federation. N.V. Beloborodova has many years of clinical experience in the field of intensive care and surgery. She studies infectious complications and sepsis. She initiated a series of interdisciplinary clinical and experimental studies based on the concept of integrating human metabolism and its microbiota. Her scientific achievements are widely known: she is the recipient of the Marie E. Coates Award "Best lecturer-scientist" Gustafsson Fund, Karolinska Institutes, Stockholm, Sweden, and the International Sepsis Forum Award, Pasteur Institute, Paris, France (2014), etc. Professor N.V. Beloborodova wrote 210 papers, five books, 10 chapters and has edited four books.

Andrey Vyacheslavovich Grechko, Ph.D., Professor, is a Corresponding Member of the Russian Academy of Sciences. He graduated from the Semashko Moscow Medical Institute (Semashko National Research Institute of Public Health) with a degree in Medicine (1998), the Clinical Department of Dermatovenerology (2000), and received a second higher education in Psychology (2009). Professor A.V. Grechko held the position of Chief Physician of the Central Clinical Hospital in Moscow. He worked as a professor at the faculty and was engaged in scientific research at the Medical University. Starting in 2013, he has been the initiator of the creation of the Federal Scientific and Clinical Center for Intensive Care and Rehabilitology, Moscow, Russian Federation, where he also serves as Director since 2015. He has many years of experience in research and teaching in various fields of medicine, is an author/co-author of more than 200 scientific publications, 13 patents, 15 medical books/chapters, including Chapter in Book «Metabolomics», IntechOpen, 2020 «Metabolomic Discovery of Microbiota Dysfunction as the Cause of Pathology».

Editors of Volume 20:
Natalia V. Beloborodova and Andrey V. Grechko
Federal Research and Clinical Center of Intensive Care Medicine
and Rehabilitology, Moscow, Russian Federation

Book Series Editor: Miroslav Blumenberg
NYU Langone Medical Center, New York, USA

Scope of the Series

Biochemistry, the study of chemical transformations occurring within living organisms, impacts all of the life sciences, from molecular crystallography and genetics, to ecology, medicine and population biology. Biochemistry studies macromolecules - proteins, nucleic acids, carbohydrates and lipids –their building blocks, structures, functions and interactions. Much of biochemistry is devoted to enzymes, proteins that catalyze chemical reactions, enzyme structures, mechanisms of action and their roles within cells. Biochemistry also studies small signaling molecules, coenzymes, inhibitors, vitamins and hormones, which play roles in the life process. Biochemical experimentation, besides coopting the methods of classical chemistry, e.g., chromatography, adopted new techniques, e.g., X-ray diffraction, electron microscopy, NMR, radioisotopes, and developed sophisticated microbial genetic tools, e.g., auxotroph mutants and their revertants, fermentation, etc. More recently, biochemistry embraced the 'big data' omics systems.

Initial biochemical studies have been exclusively analytic: dissecting, purifying and examining individual components of a biological system; in exemplary words of Efraim Racker, (1913 –1991) "Don't waste clean thinking on dirty enzymes." Today, however, biochemistry is becoming more agglomerative and comprehensive, setting out to integrate and describe fully a particular biological system. The 'big data' metabolomics can define the complement of small molecules, e.g., in a soil or biofilm sample; proteomics can distinguish all the proteins comprising e.g., serum; metagenomics can identify all the genes in a complex environment e.g., the bovine rumen. This Biochemistry Series will address both the current research on biomolecules, and the emerging trends with great promise.

Contents

Preface

The human microbiome is a collection of all microbiota that live in the human body: in the oral cavity and gastrointestinal tract, on the upper respiratory tract, in the urinary and genital tracts, in biological media (saliva, sputum, bile, feces, urine, seminal fluid, etc.). Humans are colonized by many microorganisms because the term «human microbiome» is sometimes used to refer to the collective genomes of all resident microorganisms including bacteria, archaea, fungi, protists, and viruses. The term «human microbiota» is used more often in scientific research when referring to micro-organisms that do not cause diseases under normal conditions (non-pathogenic). Some micro-organisms that colonize humans are commensal, meaning they co-exist without harming humans; others have a mutualistic relationship with their human hosts.

Interest in the role of the human microbiome is extremely high, and for a long time, it was limited by the difficulties of cultivating numerous types of microorganisms that are part of the community. Modern laboratory technologies have made it possible to fully assess not only the taxonomic composition of microbial communities but also to gain new knowledge about the functional characteristics of the relationship between the microbiota and the human body.

This book presents data from recent years confirming the huge role that the microbiome plays in the human body. The results of scientific research using state-of-the-art technologies allowed us to establish that not only normal taxonomic diversity is important for human health, but also adequate functional activity of the microbiota. The book provides data on the features of changes in the microbiota in a number of problematic diseases and discusses hypotheses about cause-and-effect relationships in the microbiome-metabolome system.

In different chapters of the book, the results of fundamental and clinical research concerning the features of the microbiome in inflammatory, critical state, other pathological conditions, and diseases (stroke, cancer, autism, allergy, psoriasis, colitis, liver diseases, etc.) are presented; much attention is paid to the interaction of the microbiota and its metabolites with drugs and natural products. The conceptual data synthesis can be a basis for the development of new approaches to treatment when one of the important therapeutic targets will be the microbiota and its key metabolites.

The problem of the human microbiome is huge and immeasurable, it goes far beyond microbiology and includes almost all of medicine. Experience shows that young scientists and doctors actively participate in scientific research, feeling the prospects of this direction for solving problems of difficult-to-treat diseases. We would like to sincerely thank the authors for the pleasure of working on this book and express the hope that it will be interesting and useful to our dear readers.

Natalia V. Beloborodova and Andrey V. Grechko
Federal Research and Clinical Center of Intensive Care Medicine and Rehabilitology,
Moscow, Russian Federation

Section 1

Microbiome Research Technologies

Chapter 1

Genomic Techniques Used to Investigate the Human Gut Microbiota

Akhlash P. Singh

Abstract

The human gut is the complex microbial ecosystem comprises more than 100 trillion microbes also known as microbiota. The gut microbiota does not only include about 400–500 types of bacterial strains, but it also contains archaea, bacteriophage, fungi, and protozoa species. In order to complete the characterization of the gut microbial community, we need the help of many culture-dependent and culture-independent genomic technologies. Recently, next-generation sequencing (NGS), mediated metagenomics that rely on 16S rRNA gene amplification, and whole-genome sequencing (WGS) have provided us deep knowledge related to important interactions such as host-microbiota and microbe-microbe interactions under various perturbation inside the gut. But, we still lack complete knowledge related to unique gene products encoded by gut meta-genome. Hence, it required the application of high-throughput "omics-based" methods to support metagenomics. Currently, a combination of high-throughput culturing and microfluidics assays is providing a new method to characterize non-amenable bacterial strains from the gut environment. The recent additions of artificial intelligence and deep learning to the area of microbiome studies have enhanced the capability of identification of thousand microbes simultaneously. Given above, it is necessary to apply new genome editing tools that can be used to design the personalized microflora which can be used to cure lifestyle-related diseases.

Keywords: culturomics, gut microbiota, human microbiome, metagenomics, metaproteomics, metabolomics, microfluidics, "multi-omics", personalized diet

1. Introduction

In the beginning of the twenty-first century, the human genome was sequenced. The main aim of this gigantic scientific effort was to identify all genes present in the human genome, also considered as the "blueprint of human life." Since then, most of the efforts are focused on the identification of all genes and annotate their functions which are responsible for genetic variation prevailed in human physiology and its association with diseases [1]. Currently, many experiments have proved that the gut microbes are more responsible than host genetics in the development of life style-related diseases. Hence, it becomes essential to investigate the crucial roles played by gut microbes in health and diseases. The human gut is a complex microbial ecosystem which is comprised of approximately 100 trillion microbes collectively known as "gut microbiota" [1]. It does not only include about 400–500 types of bacterial species but also contains archaea, bacteriophage, fungi, and protozoa species [2]. According to a rough estimation, the human gut microbiome contains almost 3.3 million genes which are 150 times more than total human genes present in the human genome. Currently, gut microflora is also considered as "gold mines" because of its commercial value in the area of biopharmaceuticals and bioactive products. In order to complete the characterization of the gut microbial community and its mysteries, we need the help of many traditional and modern genomic technologies developed in due course of time. However, the study of the human microbiome is relatively a newly emerging area in the area of human biology, thus called the "forgotten organ" in the human body.

The study of the human microbiome was started with the help of reductionist approaches such as identification and characterization of a single bacterial strain by using culture media and microscopes. Initially, only culturable bacteria could only be identified and phylogenetically classified. It is well known that more than 40% of gut microbes cannot grow outside the natural environment. Hence, both culture-dependent and culture-independent analytical methods are applied that have improved our knowledge related to human gut microbiota. Recently, next-generation sequencing (NGS) has revolutionized all areas of biological sciences including the human gut microbiome. This also supports the most traditional metagenomic technique based on 16sRNA gene amplification via polymerase chain reaction (PCR) and whole-genome sequencing (WGS) also. However, both culture-dependent and culture-independent techniques have provided the snapshot of the gut microbial community, but they are still hazy in respect of host-microbiota and microbe-microbe interactions that make stable conditions of gut microbial communities under the influence of various perturbations such as environmental factors, diets, and drugs. In the last 20 years, it becomes apparent that gut microbes add in the metabolism and contribute to strengthening the host's immune system. The human gut microbiota constitutes a metagenome that encodes an intricate network of genes, proteins, and metabolites. In order to functionally characterize human microbiome, it requires applications of many supplementary high-throughput "omics-based" methods, e.g., metaproteomics, metatranscriptomics, and metabolomics.

Recently, several labs the world over have adopted new emerging technologies to support metagenomics consequently; it amasses the terabits data in various genomic databases. To retrieve meaningful information from a large amount of multi-omics data, the application of a high level of computational and bioinformatics knowledge is required. In view of the recent explosion of data in every field, machine learning and deep learning come forward for the rescue of scientists. Therefore, different algorithms have been created, tested, and applied to huge microbiome data to identify the results of numerous microbial strains. But the next aim of all plethora of technologies is to unravel the significant contribution of gut microbiota to human biochemistry and physiology, and ultimately, this knowledge can be translated to improve human health and reduce lifestyle-related pandemic prevailed worldwide. In view of the above facts, the current chapter describes a set of analytical methods that are used to dig deep into the human gut microbial community. These methods are exploited in phylogenetic classification and functional characterization of gut microbiota.

2. A brief history of the human microbiome study

The field of the human microbiome is closely associated with microbiology; hence, its study was started in the seventeenth century. Antonie van Leeuwenhoek, who is also considered the father of microbiology, discovered oral microbes by using a simple microscope and called them "animalcules" in 1676. In the 1800s, Robert Koch developed the investigation technique for anthrax. The pioneering work of Pasteur, Koch, Escherich, and Kendall founded a strong base of microbiome research; hence, they are able to identify and count a large number of bacterial strains. In 1907, Metchnikoff proposed that lactic bacteria can ward off against harmful or putrefying bacteria from the gut [2]. Joshua Lederberg for the first time used the term "microbiome" for gut microbial community, and its relationship with the host. The microbial community can be defined as "the set of organisms (in this case, microorganisms) coexisting in the same space and time" [3].

In the beginning, only culturable bacterial species were studied, but there are a large number of microbes that are not grown inside the lab environment. That was revealed when the number of microorganisms observed by the microscope did not match with a number of microorganisms that grow on the media plate [4]. In 1970, Carl Woese suggested that ribosomal RNA genes can be used as molecular markers for bacterial classification [5]. Thus, scientists have developed the culture-independent technique based on amplification of 16S rRNA gene by PCR method and its sequencing by Sanger method. These strategies are used to classify gut microorganisms phylogenetically and then annotate their functions in a particular natural microbial ecosystem [6]. It has revolutionized the field of "microbiome research." Other culture-independent techniques, which significantly influence the taxonomic research, were the PCR, rRNA gene cloning and sequencing, fluorescence in situ hybridization (FISH), denaturing gradient gel electrophoresis (DGGE) and temperature gradient gel electrophoresis (TGGE), restriction fragment length polymorphism (RFLP), and terminal restriction fragment length polymorphism (T-RFLP). But these techniques could not reveal the metabolic and ecological functions of microorganisms. In order to ascertain the function of individual bacterial strain in the gut ecosystem, germ-free mouse models were also developed.

But due to cumbersome and time-consuming methods of traditional metagenomic techniques, new methods based on NGS have taken over the central stage to investigate the microbial communities [7]. Currently, sequencing-based techniques are used to classify numerous uncultivable microbes. Most recently, mass spectroscopy (MS) and one of its variants, i.e., matrix-assisted laser desorption/ionization time-of-flight (MALDI-TOF)-based "omics"-based high-throughput methods, have been applied to functional characterization of microbial communities [8]. These sophisticated technologies have amassed a huge amount of genomic data that needs to be annotated by computer-based systems biology approaches. The systems biology will provide a holistic picture of the microbial community inside the human gut. Currently, seven major groups such as *Firmicutes*, *Bacteroidetes*, *Proteobacteria*, *Fusobacteria*, *Verrucomicrobia*, *Cyanobacteria*, and *Actinobacteria* which constitute a major chunk of gut microbes have been recognized, but out of these two phyla, namely, *Firmicutes* and *Bacteroidetes*, include most of the gut bacteria species [9].

3. Methodology for human gut microbiome studies

Initially, culture and biochemical typing were the standard methods to identify any new bacterial species. To know more about the human gut microbes' diversity, its compositions, and relationships with various diseases, thus, many other techniques are also developed. The evolution of various methods applied to investigate the human gut microbiota is described above. Recently, significant advancements have been made in the area of sequencing-based genome technologies including metatranscriptomics, proteomics, and metagenomics, which are further supported by culturomics and computational biology for studies of human gut microbiome research. These techniques are rapid and, hence, provided a huge wealth of genomic data related to uncultured microorganisms. This helped us in the identification of new microbe species inside the gastrointestinal tract. But there are many important issues associated with the accurate and proper investigation of a gut ecosystem like sample preparation, storage, and handling from the human as well as animal subjects. In the current chapter, total techniques under three major headings (1) culture-dependent methods, (2) culture-independent genomic technologies, and (3) latest techniques are described (**Figure 1**).

3.1 Phylogenetic analysis of microbial community

3.1.1 Culture-dependent methods

In the last century, most of the microbiome studies were based on culture-based methods. Almost for the last 300 years, this approach mainly relied on main identification features like colony features, bacterial growth, and selection of some biochemical typing and microscopic investigation of culturable microbes in lab condition. In the 1980s, large numbers of gram-negative bacterial species were identified from the fecal samples [10]. Later on, many species have been identified and phylogenetically classified by using fermentation profiling or *in vitro* requirements of bacterial species. It has contributed enormously to the identification of microbial agents and given birth to a new branch, i.e., microbial ecology [11]. The culture-based method is still considered as the gold standard protocol for the identification of new species and provided a deep understanding related to the microbial world. They are a cheap and most credible method of bacterial identification. But they could not be proven completely effective against anaerobic and not amenable bacterial species. It is already given that more than 30% of bacterial species cannot be grown outside from their habitat. Moreover, gut microbiota not only includes the bacteria but also consists of bacteriophages, archaea, fungal species, and single-celled eukaryotes. Hence, we need more wide investigative approaches to cover all the microbial agents involved and contribute to the stable form of gut microbiota.

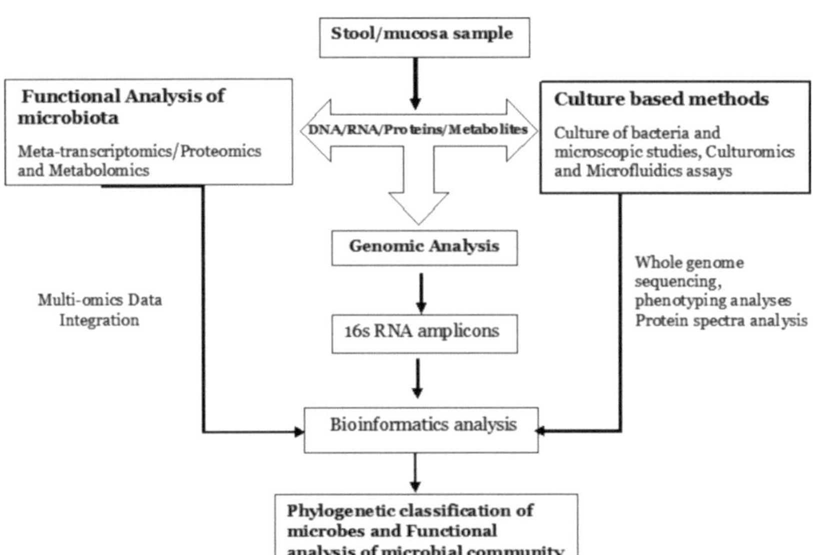

Figure 1.
Summary of techniques used to phylogenetic classification and functional characterization of the human gut microbiome.

3.1.2 Culturomics

The significance of culture-dependent methods cannot be undermined for the identification of microbes from the gut microbial community. Therefore, microbiologists have rediscovered and focused once again to revive culture-based methods by adding many sophisticated instrumentations and suitable growth media. This has allowed growing most of the unculturable bacteria that were earlier thought to be impossible in a lab environment. Hence, it will allow to know more about the functional aspects of gut microbiome that include its composition, microbial gene expression, metabolic pathways and host-bacteria relationships [12]. Actually, diverse types of favorable growing and incubation conditions are required to grow unculturable microbes that are provided by the new culturomics procedures. Currently, more than 50% of bacterial species that were earlier identified by classical 16S rRNA metagenomics could be re-identified with the help of culturomics. Simultaneously, it will also allow isolating hundreds of new bacterial species in the gut microbial ecosystem in the near future [13].

The culturomics is a multistep protocol that includes sample preparations and their diversification under different growth conditions that promote the growth of fastidious bacteria but, simultaneously, also cease the growth of few microbes. The targeted samples are subjected to further MALDI-TOF mass spectroscopy-based investigations such as a comparison of newly obtained protein spectra in recent protein databases. If the applied method could fail to establish the identification of bacteria, then the sample is processed for NGS-based 16sRNA metagenomic methods. Based on the 16sRNA gene sequencing, various toxicogenomics principles are applied to classified new species in phyla or family. Culturomics is quite an effective growth strategy particularly in microbes that are involved in mechanistic networks or intricate host-microbiome interactions. More recently, many culture techniques, for example, gel microdroplets, microculture, and microbial chips, provide very diverse growth conditions; hence, a large number of unknown microbes are able to grow [14]. Although new methods are quite helpful in the identification of new microbial species, e.g., from gut microbial ecosystem, these are also used to study human vaginal and urinary microbiota. Currently, almost 2671 new species have been identified by using culturomics ranging from commensals to pathogens, for example, 31 new bacterial species that belong to *Synergistetes* or *Deinococcus-Thermus* phyla. But, there are certain demerits like nonavailability of suitable culture media and growth conditions that allow the growth of uncultured bacteria in an artificial environment [15]. Moreover, certain bacteria grow in a highly intrigue environment inside the human gut because several microbes use common metabolites as a food and live in symbiotic and mutual interrelationship inside the gut environment.

3.1.3 Microfluidics assays

Microfluidics systems or cell on-chip offers a specific microenvironment for biochemical reactions. Microfluidics comprises numerous microchannels enshrined on the glass or polymer surface such as polydimethylsiloxane [16]. These channels are linked to each other that are based on principles of mixing, pumping, sorting, or offering biochemical environment; hence it can produce a suitable environment for microbial reactions. Recently, great advances have been made in this area; consequently, high-throughput screening, multiplexing, and automation of biochemical reactions could be achieved [17]. Microfluidics technique is also applied in the studies of gut microbiota; hence, some scientists called it gut-on-chip. With microchips, many uncultured microbes are identified because it provides specific growth environment and nutrition required for these bacterial growths, for example, microfluidics-based model (human-microbial cross talk (HuMiX)). The HuMiX provide gastrointestinal-like environment for the co-growth of human epithelial cell and obligate anaerobe *Bacteroides caccae* cells [18]. Recently developed iChip containing multiple microchambers which are further divided into hundreds of miniature multiple cells has been used to grow bacteria. This technique mainly acts by providing a selective supply of nutrients to an inoculated single bacterial cell on-chip. Another chip-based method l-tip also acts on the same principles as iChip, but it allows bacterial cell multiplication in a gel and supplies required nutrients which are essential for growth [19]. Microfluidics is the combination of gel-based methods and sophisticated instruments, for example, first we grow a single bacterial cell, then amplify its genome, and, finally, sequence its genome that helps in identifying new species [20]. Recently, TM7, bacterium, and *Sulcia muelleri* could be identified which produced very unique metabolites. By using the same method, 34 various bacterial strains are identified and phylogenetically classified.

3.2 Culture-independent methods

3.2.1 Sample collection and standardization methods

The sample preparation is a very crucial and important step of any microbial or biochemical analysis that determines the accuracy and efficacy of any simple or sophisticated analytical technique. In the human microbiome studies, there are two major types of samples, namely, stool and mucosal biopsy. However, the mucosal biopsy sample must be preferred, but their availability and handling are not easy. Ideally, stool samples must be used in conjunction with the mucosal samples [21, 22]. Several proofs of investigation have shown that there are great ambiguities prevailed between the presence of microbiota in mucosal and stool samples. Sample collection and their storage conditions also influence the final results in terms of the genetic composition of gut microbes. It has been noticed that the populations of the two most abundant gut microbial species such as *Firmicutes* to *Bacteroidetes* are affected with storage temperature in the fecal sample [23]. The sample processing methods are also held responsible for the variations in results. Hence, different consortiums associated with large-scale investigation of the gut microbiome have suggested that we must adopt the standard and calibrated protocols for sample processing [24]. Therefore, many kits are developed, for example, Qiagen QIAamp DNA Stool Mini Kit (QIAG) has significantly improved the DNA extraction and reproducibility of results from fecal samples. Moreover, researchers have also recommended other methods, namely, phenol/chloroform (PHEC), chaotropic (CHAO), and THSTI. Their comparative efficacies and performance were analyzed in terms of the final yield of DNA [26]. Currently, one more DNA/RNA Extraction Kit (TS), i.e., TianLong Stool, is also used, which mainly acts on mechanical shearing and bead beating method. It also provided good reproducible results. However, comparative studies reflect that the TS kit offers a higher quantity of nucleic acids than the other extraction kits. Conclusively, we can say that, standard protocols that are available in the form of kits that save our time and efforts of researchers [25].

3.2.2 Metagenomic analysis of microbial community

In order to overcome the drawbacks of traditional culture-based protocols, microbiologists have developed several advanced culture-independent methods to know the composition of gut microbiota. In this series, metagenomics was the first technique by which 80% of uncultured microbes are phylogenetically identified.

This culture-independent technique for microbial growth has revolutionized the area of human microflora investigations in the last two decades.

The classical techniques of metagenomics rely on the 16S ribosomal RNA (16S rRNA) gene. The 70S ribosome is the major component of prokaryotic cells and involved in protein synthesis which is highly conserved processes in all bacterial cells. The major function of 16S rRNA is the regulation of protein synthesis. During protein synthesis process, 3′ end of 16S RNA combines with the ribosomal proteins S1 and S21 involved in activation and initiation of protein synthesis. Although 16S rRNA is highly conserved among microbial species, it also contains few hypervariable regions that offer phylogenetic linkage; hence, it is also proven to be helpful in the classification of enormous microbial diversities that prevailed on earth [26]. With the development of DNA sequencing methods, 16S rRNA gene amplicons are isolated and sequenced; hence, it now becomes the most successful and prevalent culture-independent method for taxonomic classification of microorganisms. After the availability of PCR-based cloning and 16S rRNAs, gene sequencing has revolutionized the area of taxonomic classification of uncultured bacterial strains in the last two decades [27].

The metagenomic protocols include the extraction of nucleic acid from the sample followed by PCR amplification of species-specific 1500-bp-long whole 16S ribosomal RNA genes [28]. It also contains highly hypervariable regions (the V4–V5 region out of nine short hypervariable regions from V1 to V9). PCR-based amplification is carried out by using universal and specific primers, and after that, physical separation of DNA fragments are carried out on electrophoresis gels [29].

Initially, 16S ribosomal RNA gene amplification was based on cloning in a suitable host, e.g., *Escherichia coli*, and then sequencing by Sanger sequencing method. After availability of PCR based cloning of 16S ribosomal RNA gene and then, sequencing of clones (amplicons) by using any DNA sequencing method. These methods have tremendously enhanced phylogenetically the identification of the gut microbiota [30]. At that time, the pace and cost of sequencing were the great impediments that could be overcome by the advent of NGS. It is now well known that PCR-mediated protocols used for characterization of microbial diversity have certain demerits. These are attributed to PCR-based amplification of 16S rRNA gene, which is a multi-step process that introduced several ambiguities into the final results, and it became more error prone due to the PCR-based sequencing method, e.g., pyrosequencing [33]. Generally gene-specific amplifications are primer based which must be appropriate for all major taxa. Furthermore, the amplified DNA fragments can harbor mutations because of the nonspecific binding of PCR primers to template DNA strands [31].

Recently, next-generation DNA sequencing has made metagenomic and whole-genome sequencing metagenomic methods more rapid and highly sophisticated. The latest sequencing methods such as 454 pyrosequencing, Illumina, SOLiD, Ion Torrent, and single-molecule real-time (SMRT) circular consensus sequencing equipment from Pacific Biosciences [32] and Oxford Nanopore have provided more pace and deep analytic power to the analyzed gut microbiome [33]. More recently, the application of Oxford Nanopore in gut microbe analysis can overcome the abovementioned PCR-based limitations such as PCR temperature, cloning, and long and deep sequencing by MinION™ nanopore sequencing technologies.

3.2.3 Real-time PCR

It is well known that PCR is a nonquantitative technique, but its variant, real-time PCR also known as quantitative PCR (qPCR), is used for microbiome analysis particularly for phylogenetic analysis. It can be used quantitatively and semiquantitatively depending upon the applications; qPCR can quantify the amount of DNA in the stool or gut mucosa samples. In this technique, fluorescent probes or dye molecules are used that intercalate between the double strand of DNA molecules or 16 s RNA amplicons. These probes send a strong signal, and its intensity is directly proportional to the amount of DNA sample present. Sometimes sequence-specific oligonucleotide probes are linked with molecular markers or complementary DNA sequence [34]. The primers designing is a crucial step in the RT-PCR technique; therefore, primers must be specific for all bacterial phyla or taxa or species present in a sample [35]. Real-time PCR has been used to investigate the state of the ecological environment in normal and obese persons [36]. Quantitative PCR technique is also used solely or in combination with other gel and non-gel-based techniques. This combination of protocols is used to understand the functional microbial diversity of gut microbiota in the patient of age and effect of antibiotics on gut microbes [37], for example, DGGE and qPCR.

Real-time PCR-based methods are suitable for the prediction of accurate phylogenetic analysis. The appropriate primers provide great help to know the composition of a microbial community and microbial load. The protocol is simple to complex, and all chemicals and consumables are easily available in laboratories. But, this is also suffering due to PCR biases, which percolate at each step of the protocol. Quantitative PCR cannot be used to detect new bacterial strains in the gut microbiota without prior information of primers or probe.

3.2.4 Genetic fingerprinting of gut microbiota

There are many culture-independent methods which mainly rely on gel-based separation and hybridization of 16sRNA sequences with the probe, for example, T-RFLP , DGGE, TGGE, and a combination of FISH and flow cytometry [38]. These methods are also known as fingerprinting methods have been used to investigate microbial diversity. In the last two decades, fingerprinting methods have offered more information related to the composition of gut microflora. This group of techniques does not provide information about the phylogenetic compositions of the gut ecosystem. But the disturbance in the composition of gut microbiome, which is also known as "gut dysbioses," caused by various environmental perturbations, including foreign bacterial species and antibiotics, could be investigated in the case of humans [39].

3.2.5 Denaturing gradient gel electrophoresis

It is the most widely used method built on the separation of 16S rRNA gene amplicons on polyacrylamide gel electrophoresis from the complex mixture of DNA fragments that have the same length but different nucleotides sequence [40]. The electrophoretic separation of DNA fragments is influenced by the gel gradient generally produced by denaturant agents, for example, urea and/or formamide. Actually, when the current passes through the electrophoresis gel, 16S rRNA gene amplicons/DNA fragments get separated at various positions on gel according to their molecular weight in linear order, and it continues till their complete denaturation. Consequently, a heterogeneous mixture of DNA sequences is separated in the form of bands on the gel due to their compositions and denatured gradient present in the gel. DGGE is a semiquantitative technique and practiced in the comparison of two different types of microbial communities, i.e., from a healthy or diseased person. The technique is fast and can be used for the separation of multiple samples in single experiments [41]. The main disadvantage of DGGE is that the final results are influenced by PCR-originated bias and not suitable for direct identification of new strains without the availability of a compatible probe.

3.2.6 Temperature gradient gel electrophoresis

It is well known that the DNA sequence influenced the value of the melting temperature (Tm) of a fragment. The high GC content is mainly responsible for high Tm, while the high AT content, for lesser Tm. That can be attributed to the fact that base pairing between G and C contains three hydrogen bonds, while A and T form two hydrogen bonds. Therefore, GC base pairing is more stable than AT in a DNA fragment. In the case of TGGE, denaturant agents are replaced with a temperature gradient. The final results of TGGE protocol mainly depend on amplicon stability and melting behavior, which are determined by GC content. Therefore, when current is passed through the slab gel, intact DNA strands get separated under the influence of temperature gradient inside the gel, but simultaneously, their movements are halted. Consequently, a banding pattern is produced under the influence of the temperature gradient; it is also known as fingerprinting or TGGE [42]. The technique of TGGE is fast and semiquantitative, but like DGGE, its results are also influenced by PCR predispositions. TGGE is not suitable for direct identification of microbes and phylogenetic analysis in absence of sequence-based suitable probes or appropriate hybridization processes.

3.2.7 Terminal restriction fragment length polymorphism assay

RFLP is a classical molecular biological technique used for genetic fingerprinting in the case of animals and plant samples. Its variant T-RFLP is applied to compare the microbial communities and the microbial diversities of gut microbiota. In the process of T-RFLP technique, 16sRNA gene amplicons are isolated from different stool samples and then amplified by PCR. Next, 16sRNA gene amplicons are cut by using different types of restriction enzymes that produced restriction fragments of varying lengths following the isolation of the electrophoresis gel. So that due to different length/M. wt, restriction fragments move to different distances on gel, thus producing a banding pattern. Being fluorescent, each terminal fragment can be identified, whereby each band represents an individual species in the gut community. T-RFLP is used in the comparison of two ecological communities [43]; it is a fast and cheap technique, but not suitable for direct phylogenetic analysis of bacterial strains. Moreover, incompatibility between primer and target genomic DNA influences the T-RFLP results [44]; therefore, it can underrepresent the crucial species, for example, *Lactobacillus* and *Actinobacteria*.

3.2.8 Probe hybridization-based methods

Probe hybridization techniques are mainly used for species identification and their quantification in particular samples. These methods depend on the complementarity between specific oligonucleotide probes and specific target DNA sequences in the bacterial genome. Two major techniques, namely, FISH and DNA microarrays, are included in this class of probe hybridization-based methods which are mainly used in phylogenetic identification and quantification of species living in the microbial ecosystem.

3.2.8.1 Fluorescence in situ hybridization

Basically, FISH is a cytogenetic technique that is applied to pinpoint a specific DNA sequence on the chromosomal landscape by using a suitable fluorescent probe. But, it is also widely used in gut microbiome studies, also known as bacterial

FISH. In the studies of microbial communities, the 16S rRNA gene amplicons are prepared and denatured in a solution. After that, both fluorescent probe and DNA strands are also added in the hybridization solution. In order to allow maximum hybridization process, some cross-linking agents like aldehyde or any precipitating agent (methanol) are also added and incubated in the reaction mixture and kept at 65–75°C for 12 h [45]. After ensuring that the hybridization process is completed, the intensity of fluorescence is measured by using suitable laser available fitted in the flow cytometry instrument. The combination of FISH and flow cytometry is a sort of high-throughput method used in the genome comparison of two different species in the gut sample [46]. The FISH technique is efficiently applied to compare two types of microbial communities such as breast- and formula-fed newborns, and two different species *Bifidobacterium* and *Atopobium* are identified [47]. The merits of this method are that it is semiquantitative and rapid. Due to the availability of diverse probes for specific phyla or species, FISH can be widely used in microbiome studies. But the technique completely failed to identify de novo identification of a bacterial strain. Some researchers have used FISH to estimate the time of sample stability and change in their species compositions with the passage of time and storage conditions.

3.2.8.2 DNA microarrays

DNA microarray technology or DNA chip method is widely applied to learn more about the microbial ecosystem, particularly in gut microbiota. The component of the DNA microarray is a small chip containing a large number of microscopic spots on a solid surface which are used to immobilize fluorescent probes. DNA spots hold pico-level DNA, which is sufficient for hybridization process of a small part of a gene or its regulatory element with cDNA already immobilized on a DNA chip under suitable reaction environments. The microarray protocol includes the following: firstly, the 16S rRNA amplicon or extracted DNA from the samples is processed to make them fluorescent. Secondly, oligonucleotide probes are spotted and immobilized on the surface of the microarray chip [48]. Finally, hybridization is allowed between 16S rRNA amplicons and fluorescent probes. The fluorescence intensity after complete hybridization is quantified by using a laser. The microarray can identify the expression of hundreds of genes in a single experiment. The effect of *C. difficile* infection and its successful cure by fecal microbiota transplantation (FMT) is studied by microarray [49]. This method is quite fast and rapid and offers a high-throughput method for phylogenetic analysis of gut microbiota. It requires a very small amount of DNA for accurate analysis. The most noticed demerit of a microarray experiment is the possibility of cross hybridization, i.e., binding of multiple oligonucleotide probes to a single DNA fragment. In the absence of the probe, a microarray cannot identify a new bacterial species.

4. Functional analysis of the microbial community

4.1 Next-generation sequencing-based methods

Before the advent of NGS, the Sanger sequencing method was the only protocol available to read DNA sequence or full-length 16sRNA gene amplicons. Sanger method was based on the DNA replication process and capillary electrophoresis. In this procedure, all components required for DNA synthesis, i.e., enzyme DNA polymerase, primers for 16sRNA gene, four types of deoxynucleotides (dATP, dGTP, dCTP, dTTP), and four types of fluorescent chain terminators (dideoxynucleotides: ddATP, ddGTP, ddCTP, ddTTP), are added to single-stranded template DNA and initiate the DNA synthesis process. Consequently, new DNA fragments of various lengths are synthesized with corresponding fluorescent chain terminators which stop further elongation of strands. Hence, randomly terminated DNA fragments are produced that are isolated with capillary gel electrophoresis. On the slab gel, four types of fluorescent dideoxynucleotides fragments can be read by a suitable laser scanning method on the basis of light emitted by them [50]. Therefore, a nucleotide sequence of 16sRNA gene amplicon can be inferred that can be searched in a large number of databases. There are many databases used for the 16sRNA gene amplicon, for example, GenBank and ribosomal RNA gene bank.

Sanger sequencing method not only supports the traditional metagenomic experiments but also supplemented to DGGE, TGGE, and T-RFLP methods as well as whole-genome sequencing metagenomics. The protocol includes the combination of gel and DNA sequencing based methods. In this, the isolated DNA bands from DGGE, TGGE, and T-RFLP gels are removed and sequenced by Sanger's sequencing methods. But in the case of scarcity of DNA, in a particular band, it can be further amplified by PCR and then sequenced. Sanger's sequencing method is most suitable to quantify and carry out phylogenetic identifications of the gut microbiota. Sanger method belongs to first-generation sequencing (FGS) technology, being the most important tool, and is also used for first human genome sequencing. This method is still considered as the gold standard method for long-read sequencing up to 500 nucleotides which are highly essential for genome assemblies. The main disadvantage associated with the Sanger method is its high cost and time-consuming nature.

In the beginning of the twenty-first century, many high-throughput methods of DNA sequencing were developed, for example, pyrosequencing which is a PCR-based massively parallel sequencing platform like Roche/454 pyrosequencing exploited for investigation of gut microbiota. It provided huge genomic data related to human microbiome analyses. Pyrosequencing technique is cheap and high-throughput and requires a small amount of DNA, but short read is a major limitation of the method and unsuitable for comparisons between species within the genus and bioinformatics analysis [51]. Parallelly, other next-generation sequencing platforms for DNA sequencing are also developed such as Illumina, SOLiD, Ion Torrent, and single-molecule real-time circular consensus sequencing equipment from Pacific Biosciences and Oxford Nanopore [52]. These technologies have to make microbiome analysis very fast and easy and amass the genomic data for phylogenetic analysis. NGS has provided great speed and accuracy to culture-independent methods used for the study of the functional diversity of microflora. Recently, MinION™ nanopore sequencing technologies used PCR-independent methods; hence, this is free from PCR-based cloning biases, such as amplification temperatures and biased primers sequences. Simultaneously, nanopore sequencing methods offer long reads, which are more suitable for genome assemblies. The above said NGS methods are applied to sequenced cloned amplicons or total community DNA [53]. These methods allow us to investigate gut microbiota qualitatively as well as quantitatively which is influenced by various perturbations, e.g., environmental factors, perturbation, and diets.

NGS is not only useful in phylogenetic classification but also helps in the functional analysis of microbial communities. Therefore, several supplementary technologies also emerged which can differentiate between microbial species in an ecosystem. But it requires analysis of different molecular signatures like DNA, RNAs, proteins, and metabolites generally produced by microbial communities. NGS provided the basic foundation for many omics-based methods, for example, metatranscriptomics, metaproteomics, and metabolomics, which have helped us in the functional analysis of metagenome represented by a whole microbial community [54]. These methods offered a huge amount of genomic data stored in different databases that can be integrated with the help of bioinformatics tools.

4.1.1 Metatranscriptomics

In fact, transcriptomics is the analysis of the whole gamut of RNA molecules expressed by a particular cell. There are many RNA molecules including mRNA, rRNA, tRNA, and other noncoding RNA transcribed in a microbial ecosystem which play an important role in the gene expression or metagenome expression in the case of the microbial community. Traditionally, the transcriptomics analysis is carried out by measuring the level of RNA expression by using cDNA-based microarray chip. To study microbial communities, thousands of fluorescent probes were required to be immobilized on the microarray chip surface. Actually, metatranscriptomics is the studies of RNA molecules encoded by a metagenome present in a local ecosystem, for example, gut microbiota. Recently, metatranscriptome is studied with the help of the RNA-seq method; this technique is extremely suitable to confirm the gene expression of complete metagenome in the sample which provides the basic data for proteomics and metabolomics [55]. Metatranscriptomics is highly sensitive methods which can even differentiate between dead and live bacterial cell present in a sample. The major drawback of the method is its high cost and it requires great care during the design and execution of experiments because of the momentary stability of mRNA and its contaminations. There are several demerits associated with this method, for example, less amount of mRNA in bacteria, and hence, it creates an experimental problem. Recently, metatranscriptomics methods have been used to identify the pathway of carbohydrate metabolism and energy extraction and physiological functions regulated by a metagenome [56].

4.1.2 Metaproteomics

The proteome is the complete protein complement expressed by a cell or tissue at a particular moment, and the study of the proteome is known as "proteomics." The metaproteomics or community proteomics is the variant of proteomics in the sense that it is the protein complement expressed by a metagenome from a microbial community. Currently, a small number of reports are available on gut community meta-proteomics that is attributed to the small amount of proteins available in the sample, and its detection makes it further a less applied method in comparison to metagenomics and metatranscriptome. There are still lacking standardized protocols related to protein extraction and its downstream processing. The detection of low abundant proteins in the sample is still a challenge. Moreover, its high cost, time-consuming, and labor-intensive nature further restricted its applications. But many labs have applied metaproteomics in the study of functional analysis of host-microbiome interactions and proteins expressed by gut metagenome. There are two types of proteomics methods, i.e., gel-dependent and gel-independent methods. First, the category of protocols includes the combination of 2D gel electrophoresis, mass spectroscopy, and various bioinformatics tools. Second, categories, namely, shotgun proteomics, mainly depend on most expensive and more sophisticated instruments like two-dimensional liquid chromatography (LC) coupled with nano-spray tandem mass spectrometry (nano 2D LC–MS/MS) and powerful bioinformatics data analysis pipeline. Both types of technologies have provided large-scale protein analysis data in the case of the human gut proteome [57]. Currently, metaproteomics methods are applied to analyze the effect of dietary components, e.g., resistant starch on protein expression, enzymes, and composition of microbes involved in starch metabolism inside the gut. This technique is useful to investigate the ratio of two important bacterial species *Firmicutes* to *Bacteroidetes* inside gut microbiota [58].

4.1.3 Metabolomics

Metabolites are the final outcome of the gene expression process; they are highly unique in the case of the gut microbiota. Large numbers of metabolites are produced by gut microbiota, which can act as pharmaceutical agents or bioactive products. The metabolomics is a high-throughput omics-based method that mainly deals with the identification and quantification of total metabolites produced in a cell, tissue, and organ which are also called the metabolome. The "meta-metabolome" is the whole complement of metabolites and is produced by a specific microbial community. The analysis of meta-metabolomics requires a set of very sophisticated tools and techniques like matrix-assisted laser desorption/ionization time-of-flight, secondary ion mass spectrometry (SIMS), and Fourier transform ion cyclotron resonance MS that are used for metabolome analysis [59]. The complete annotation of the metabolome produced by a metagenome will help us to understand the physiology and functionality of a microbial community. Inside the human gut, fermentation of short-chain fatty acid is carried out by specific bacteria and produced many types of metabolites that participate in host metabolism and influence the physiology of both host-microbial communities inside the gut. The metabolome analysis offered the investigation of functional gene products in a sample that is helpful in functional analysis of microbes present a microbial niche. Currently, many unique metabolites are identified that are produced by gut microbiota.

4.1.4 Bioinformatics and multi-omics data integration

In the last two decades, bioinformatics has provided much needed help to annotate the complex genome sequences and metagenomic data. The microbial bioinformatics offers help to understand microbial agents of the microbial ecosystem and their mutual and host-microbes interactions. Recently, community-based bioinformatics platforms and pipelines are developed like Mothur and QIIME which help in downstreaming of high-throughput genome sequencing data of variable regions of bacterial 16S ribosomal genes or amplicons. These platforms also help in data analysis and visualization of gut microbiome composition. The high-throughput method like shotgun sequencing and WGS metagenomics produced a huge amount of data, and its annotation is a great challenge in the field of microbiome analysis [60].

In order to know the functions of a particular microbial community, it requires integrating data from other studies such as metatranscriptomics sequencing, metagenomics, metatranscriptomics, metaproteomics, metabolomics, and other techniques. The integration of data provides holistic knowledge of a gut community in terms of its structure and functions [61]. For example, any perturbation such as antibiotics or heavy metal toxicities leads to the change in gut microbial community that can be studied at the level of metabolite production and protein expression. Multi-omics data integration is the uphill task and requires a highly advanced level of computational skill, but current few tools have been developed, e.g., XCMS is a new web-based tool that integrates transcriptome, proteome, and metabolome data [62]. The new systems-level integration can also provide valuable insights, especially when they are combined with community surveys and metagenomics (**Table 1**).

Technique	Basis of techniques	Advantage	Disadvantage
Method for phylogenetic classifications			
Culture-dependent methods			
Culture of bacteria and microscopic studies	Colony features, microscopic and biochemical studies	Low cost	Not suitable for microbiota studies
Culturomics	Culture of microbes and MALDI-TOF mass spectroscopy-based investigations	Appropriate for uncultured microbes	Extremely costly
Microfluidics assays	Microchips based on biochemical reactions	Co-culture of microbes	Need high technical knowledge
Culture-independent			
Quantitative PCR	Fluorescent dyes bind with 16S rRNA gene and quantification of DNA	Highly suitable for phylogenetic classification	Not suitable to identify new bacterial species and biased due to PCR steps
DGGE/TGGE	Separation of 16S rRNA amplicons on electrophoresis based on DNA denaturants and temperature gradients	Fast, less-expensive, and semiquantitative	Results also affected by PCR biases
T-RFLP	Fragmentation of 16S rRNA amplicons by one or more restriction enzymes followed by electrophoretic isolation	Fast and semiquantitative	Not suitable for phylogenetic identification, results are affected with PCR biases
FISH (fluorescence in situ hybridization)	The 16S rRNA amplicon-specific fluorescent probes and flow cytometry	Used for phylogenetic analysis	Unable to identify a new bacterial species. Free from PCR-based bias
DNA microarray	Fluorescent probes immobilized on DNA chip hybridized with 16S rRNA gene. Fluorescence intensity is measured by laser	Phylogenetic identification is possible, high-throughput method, a semiquantitative method which is very fast	Possibility of cross hybridization, PCR biases, detect low-level species in gut microbiota
Cloning of 16sRNA gene (*classical metagenomics*)	Amplification of 16sRNA gene by PCR- and Sanger-based sequencing by capillary electrophoresis	Highly suitable for phylogenetic classifications and microbiota composition	Affected with the PCR/cloning bias, time-consuming, and extremely expensive
Direct sequencing of 16sRNA amplicon (*modern metagenomics*)	Sequencing of 16S rRNA amplicons by fast NGS methods, e.g., 454 pyrosequencing, Illumina, SOLiD, single-molecule real-time, Pacific Biosciences and nanopore sequencing methods	Cheap, fast, suitable for phylogenetic identification of unknown microbes	PCR biases, expensive, laborious, and computer intensive
Whole-genome sequencing of bacterial species	Sequencing of the whole genome by NGS-based methods	Suitable for phylogenetic identification of new species	Expensive and computer intensive

Technique	Basis of techniques	Advantage	Disadvantage
Shotgun cloning of microbiome genome/ metagenome	Random shearing of genome. Then assemble genomes on the basis of overlapping sequences by bioinformatics methods	Useful for phylogenetic identification of new species and suitable for microbiome studies	Method is costly and not suitable for phylogenetic classification of a new bacterial species
Method for functional analysis			
Multi-omics methods			
Metatranscriptomics	Sequenced RNA molecules encoded by a metagenome through NGS-based RNA-seq methods	Can identify the metabolism encoded by metagenome	Expensive and requires technical knowledge to conduct experiments
Metaproteomics	Detection of all proteins encoded by metagenome by applying nano 2D LC–MS/MS	Can identify the unique proteins and enzymes encoded by metagenome	Difficult to protein extraction and its downstream processing
Metabolomics	Detection of all metabolites encoded by metagenome using MALDI-TOF, SIMS, and Fourier transform ion cyclotron resonance MS	The method can be used to identify noble metabolites and metabolic pathways imparted by a microbial community	Highly expensive and sophisticated, lack of standard protocols so far
Bioinformatics	Various web-based data analysis pipelines/ platforms are developed QIIME and XCMS	Integration of omics-based data, it provides holistic knowledge about gut microbiome	Need high level of computational skill

Table 1.
Summary of various techniques used for phylogenetic classification and functional characterization of the human gut microbiome.

5. New advancements

5.1 Machine learning

The advancements made in the area of NGS also coincide with machine learning in the last two decades. Machine learning, a branch of artificial intelligence, is based on computational and statistical principles and is recently applied to various fields of genomics including microbiome genomics. Machine learning deals with the development and testing of algorithms to identify, classify, and forecast patterns that emerged from a huge data set [63]. The gut microbial community is comprised of trillion of microbes which further affected various types of factors such as diet, drugs, age, environment, and even lifestyles. To extract the information from such an intricate system cannot be carried out by humans but rather require machine intervention. The machine learning methods such as deep learning and neural network are used to predict severity and susceptibility gingivitis on the basis of the oral microbiome. The two most important machine learning algorithms, random forest and SourceTracker, are applied to know the effect of antibiotics on the genomic and metagenomic studies [64]. In the near future, machine learning can be used to know the host-trait prediction.

5.2 Genome editing/synthetic biology of microbial community

Genome sequencing data of thousands of bacteria are now available in various databases. Currently, many types of genome editing tools are available to manipulate the genome of animals and plants including microbial genomes. Many scientists have exploited these tools in the manipulation of gut microbiota so that desirable genetic changes can be brought into the metagenomes. The most widely used genome editing tool CRISPR-Cas systems also called clustered regularly interspaced short palindromic repeats (CRISPRs) and CRISPR-associated (Cas) proteins are present in the microbes which are mainly responsible for adaptive immunity for prokaryote cells. CRISPR-Cas systems comprise combinations of short DNA sequences called spacers that guide Cas proteins to cleave foreign DNA. So far, CRISPR-Cas systems are the most widely studied and applied method used for genetic manipulation. There are several types of a spacer or genome editing CRISPR-Cas systems, for example, Cas9, CasX, and CRISPR-CasY, that can be used to manipulate genomic content of gut microorganisms. Class 2 CRISPR-Cas systems are streamlined versions, in which a single RNA-bound Cas protein recognizes and cleaves target sequences. Actually, components of Class 2 CRISPR-Cas systems are studied, and assembly from its components in vitro system has revolutionized the field of synthetic biology.

The gut microbiota also comprises a microorganism, for example, single-cell eukaryotes, bacteria, fungus, and bacteriophages. They live in the gut in a very harmonious manner with trillions of bacteria in a natural environment, hence, well adapted to the local environment. Therefore, researchers are embarking on the idea that gut symbionts can be potential agents or vectors for genetic manipulation of gut microbial communities. The new genome editing tools are used to genetically reprogram gut communities under synthetic biology [65]. CRISPR-Cas systems have been exploited to modification of gene expression, change of the production of metabolites, biocatalyst, and protein production that can act as better microbiome modulators. Moreover, genome editing tools will prove extremely helpful in the functional characterization of gut microbiota. Current genome editing tools have offered opportunities in the investigation of intricate relationships between members of the microbiome and host and have opened new avenues for the development of pharmaceutical agents that target the microbiome. But still many demerits are also linked with genome editing tools including their off-targets and inability to introduce exogenous DNA into the metagenome [66]. Moreover, many bacteria particularly unculturable are naturally ill-adapted to transformation methods such as electroporation, conjugation, or transduction in lab conditions.

6. Summary and future prospectus

The gut microbiome is an unexploited huge wealth of microbes that synthesized the valuable and unique metabolites to be used for pharmaceutical industries and the preparation of functional foods. Additionally, metabolites produced by the gut microbiome also contribute in maintaining the health and immunity of the host. In order to exploit microbiome's wealth, we need to apply appropriate and suitable analytical techniques in a highly systematic manner to dig out unique biomolecules. The gut microbial community contains trillions of microbes that make it highly complex. It carried out thousands of metabolic and biochemical reactions in the natural environment. Hence, investigating gut microbiota requires new culturomics methods because of a large number of microbes not able to grow in an artificial environment. Currently, data generated by high-throughput sequencing contain a wealth of information and must be analyzed by using advanced tools and techniques of bioinformatics and microbiology techniques.

Now the picture of the human gut microbiome is available but still hazy in terms of how microbes impact their host and other microbes living in the gut microbial community. The NGS has revolutionized every field of biological sciences including human microbiome research. It not only sequenced thousands of genome of microorganisms but also helped to emerge many supplementary technologies which are very significant in the functional investigation of the microbial community. Therefore, the advent of modern "omics-based" high-throughput methodologies will help in the identification and characterization of previously unknown microbial strains and modulation mechanisms of the gut ecosystem. But the huge data generation by the omics-based methodologies is a great challenge which needs to be dealt with the development of new bioinformatics tools and techniques. Simultaneously, methods of big data analysis also need to be designed like machine learning and deep learning that will certainly help us in the study of microbial communities.

The availability of cheap and sufficient raw data has opened new avenues. In the near future, gut microbiota can be used as biomarkers and can be personalized to microflora on the line of personalized diet and personalized genomics. Moreover, the recent development of genomic editing tools can manipulate the microbial community under the techniques of synthetic biology. Hence we can cure lifestyle-related diseases such as obesity, cancer, and diabetes by positive manipulation in the composition of gut microbiota.

Acknowledgements

I would like to thank to the Department of Biochemistry, GGDSD College (Panjab University), India, for providing me all the facilities and support for preparation of this manuscript.

Conflict of interest

The authors declare no conflict of interest.

Author details

Akhlash P. Singh
Genomics and Proteomics Lab, Department of Biochemistry, GGDSD College, Panjab University, Chandigarh, India

*Address all correspondence to: akhlash@ggdsd.ac.in; akhlash@gmail.com

References

[1] Grice EA, Segre JA. The human microbiome: Our second genome. Annual Review of Genomics and Human Genetics. 2012;**13**:151-170

[2] Savage DC. Microbial ecology of the gastrointestinal tract. Annual Review of Microbiology. 1977;**31**(1):107-133

[3] Steinhoff U. Who controls the crowd? New findings and old questions about the intestinal microflora. Immunology Letters. 2005;**99**(1):12-16

[4] Staley JT, Konopka A. Measurement of in situ activities of nonphotosynthetic microorganisms in aquatic and terrestrial habitats. Annual review of microbiology. Oct 1985;**39**(1):321-346

[5] Woese CR, Fox GE. Phylogenetic structure of the prokaryotic domain: The primary kingdoms. Proceedings of the National Academy of Sciences. 197;**74**(11):5088-5090

[6] Staley JT, Konopka A. Measurement of in situ activities of nonphotosynthetic microorganisms in aquatic and terrestrial habitats. Annual Review of Microbiology. 1985;**39**(1):321-346

[7] Woese CR, Fox GE. Phylogenetic structure of the prokaryotic domain: The primary kingdoms. Proceedings of the National Academy of Sciences. 1977;**74**(11):5088-5090

[8] Zoetendal EG, Vaughan EE, De Vos WM. A microbial world within us. Molecular Microbiology. 2006;**59**(6):1639-1650

[9] Amann RI, Ludwig W, Schleifer KH. Phylogenetic identification and in situ detection of individual microbial cells without cultivation. Microbiological Reviews. 1995;**59**:143-169

[10] Savage DC. Microbial biota of the human intestine: A tribute to some pioneering scientists. Current Issues in Intestinal Microbiology. 2001;**2**:1-15

[11] Backhed F, Ley RE, Sonnenburg JL, Peterson DA, Gordon JI. Host-bacterial mutualism in the human intestine. Science. 2005;**307**(5717):1915-1920

[12] Blevins SM, Bronze MS. Robert Koch and the 'golden age' of bacteriology. International Journal of Infectious Diseases. 2010;**14**(9):e744-e751

[13] Ackert L. Sergei Vinogradskii and the Cycle of Life: From the Thermodynamics of Life to Ecological Microbiology, 1850-1950. Vol. 34. Netherlands: Springer Science & Business Media; 2012. p. 194

[14] Goodman AL et al. Extensive personal human gut microbiota culture collections characterized and manipulated in gnotobiotic mice. Proceedings of the National Academy of Sciences of the United States of America. 2011;**108**:6252-6257

[15] Lagier JC, Dubourg G, Million M, Cadoret F, Bilen M, Fenollar F, et al. Culturing the human microbiota and culturomics. Nature Reviews Microbiology. 2018;**16**(9):540-550

[16] Ma L, Kim J, Hatzenpichler R, Karymov MA, Hubert N, Hanan IM, et al. Gene-targeted microfluidic cultivation validated by isolation of a gut bacterium listed in human microbiome project's most wanted taxa. Proceedings of the National Academy of Sciences. 2014;**111**(27):9768-9773

[17] Shah P, Fritz JV, Glaab E, Desai MS, Greenhalgh K, Frachet A, et al. A microfluidics-based in vitro model of the gastrointestinal human–microbe interface. Nature Communications. 2016;7:11535

[18] Nichols D, Cahoon N, Trakhtenberg EM, Pham L, Mehta A, et al. Use of ichip for high throughput in situ cultivation of "uncultivable" microbial species. Applied and Environmental Microbiology. 2010;**76**:2445-245

[19] Jung D, Seo EY, Epstein SS, Joung Y, Han J, Parfenova VV, et al. Application of a new cultivation technology, I-tip, for studying microbial diversity in freshwater sponges of Lake Baikal, Russia. FEMS Microbiology Ecology. 2014;**90**(2):417-423

[20] Arnold JW, Roach J, Azcarate-Peril MA. Emerging technologies for gut microbiome research. Trends in Microbiology. 2016;**24**(11):887-901

[21] Zoetendal EG, von Wright A, Vilpponen-Salmela T, Ben-Amor K, Akkermans AD, de Vos WM. Mucosa-associated bacteria in the human gastrointestinal tract are uniformly distributed along the colon and differ from the community recovered from feces. Applied and Environmental Microbiology. 2002;**68**(7):3401-3407

[22] Tong M, Jacobs JP, McHardy IH, Braun J. Sampling of intestinal microbiota and targeted amplification of bacterial 16S rRNA genes for microbial ecologic analysis. Current Protocols in Immunology. 2014;**107**(1):7-41

[23] Stinson LF, Keelan JA, Payne MS. Comparison of meconium DNA extraction methods for use in microbiome studies. Frontiers in Microbiology. 2018;**9**:270

[24] Bag S, Saha B, Mehta O, Anbumani D, Kumar N, Dayal M, et al. An improved method for high quality metagenomics DNA extraction from human and environmental samples. Scientific Reports. 2016;**6**:26775

[25] Lim MY, Song EJ, Kim SH, Lee J, Nam YD. Comparison of DNA extraction methods for human gut microbial community profiling. Systematic and Applied Microbiology. 2018;**41**(2):151-157

[26] Eckburg PB, Bik EM, Bernstein CN, Purdom E, Dethlefsen L, Sargent M, et al. Diversity of the human intestinal microbial flora. Science. 2005;**308**(5728):1635-1638

[27] Kolbert CP, Persing DH. Ribosomal DNA sequencing as a tool for identification of bacterial pathogens. Current Opinion in Microbiology. 1999;**2**(3):299-305

[28] Olsen GJ, Lane DJ, Giovannoni SJ, Pace NR, Stahl DA. Microbial ecology and evolution: A ribosomal RNA approach. Annual Reviews in Microbiology. 1986;**40**(1):337-365

[29] Lane DJ, Pace B, Olsen GJ, Stahl DA, Sogin ML, Pace NR. Rapid determination of 16S ribosomal RNA sequences for phylogenetic analyses. Proceedings of the National Academy of Sciences. 1985;**82**(20):6955-6959

[30] Rajendhran J, Gunasekaran P. Microbial phylogeny and diversity: Small subunit ribosomal RNA sequence analysis and beyond. Microbiological Research. 2011;**166**(2):99-110

[31] Andoh A, Benno Y, Kanauchi O, Fujiyama Y. Recent advances in molecular approaches to gut microbiota in inflammatory bowel disease. Current Pharmaceutical Design. 2009;**15**(18):2066-2073

[32] Lu T, Stroot PG, Oerther DB. Reverse transcription of 16S rRNA to monitor ribosome-synthesizing bacterial populations in the environment. Applied and Environmental Microbiology. 2009;**75**(13):4589-4598

[33] Nicholls SM, Quick JC, Tang S, Loman NJ. Ultra-deep, long-read

nanopore sequencing of mock microbial community standards. Gigascience. 2019;**8**(5):giz043

[34] Riddle MS, Connor BA. The traveling microbiome. Current Infectious Disease Reports. 2016;**18**(9):29

[35] Carey CM, Kirk JL, Ojha S, Kostrzynska M. Current and future uses of real-time polymerase chain reaction and microarrays in the study of intestinal microbiota, and probiotic use and effectiveness. Canadian Journal of Microbiology. 2007;**53**(5):537-550

[36] Kieler IN, Mølbak L, Hansen LL, Hermann-Bank ML, Bjornvad CR. Overweight and the feline gut microbiome—A pilot study. Journal of Animal Physiology and Animal Nutrition. 2016;**100**(3):478-484

[37] Zwielehner J, Liszt K, Handschur M, Lassl C, Lapin A, Haslberger AG. Combined PCR-DGGE fingerprinting and quantitative-PCR indicates shifts in fecal population sizes and diversity of Bacteroides, bifidobacteria and Clostridium cluster IV in institutionalized elderly. Experimental Gerontology. 2009;**44**(6-7):440-446

[38] Feng XW, Ding WP, Xiong LY, Guo L, Sun JM, Xiao P. Recent advancements in intestinal microbiota analyses: A review for non-microbiologists. Current Medical Science. 2018;**38**(6):949-961

[39] Lange K, Buerger M, Stallmach A, Bruns T. Effects of antibiotics on gut microbiota. Digestive Diseases. 2016;**34**(3):260-268

[40] Muyzer G. DGGE/TGGE a method for identifying genes from natural ecosystems. Current Opinion in Microbiology. 1999;**2**(3):317-322

[41] Noor SO, Ridgway K, Scovell L, Kemsley EK, Lund EK, Jamieson C, et al. Ulcerative colitis and irritable bowel patients exhibit distinct abnormalities of the gut microbiota. BMC Gastroenterology. 2010;**10**(1):134

[42] Fischer SG, Lerman LS. Separation of random fragments of DNA according to properties of their sequences. Proceedings of the National Academy of Sciences. 1980;**77**(8):4420-4424

[43] Li F, Hullar MA, Lampe JW. Optimization of terminal restriction fragment polymorphism (TRFLP) analysis of human gut microbiota. Journal of Microbiological Methods. 2007;**68**(2):303-311

[44] Polz MF, Cavanaugh CM. Bias in template-to-product ratios in multitemplate PCR. Applied and Environmental Microbiology. 1998;**64**(10):3724-3730

[45] Lukumbuzya M, Schmid M, Pjevac P, Daims H. A multicolor fluorescence in situ hybridization approach using an extended set of fluorophores to visualize microorganisms. Frontiers in Microbiology. 2019;**10**:1383

[46] Swidsinski A, Loening-Baucke V, Vaneechoutte M, Doerffel Y. Active Crohn's disease and ulcerative colitis can be specifically diagnosed and monitored based on the biostructure of the fecal flora. Inflammatory Bowel Diseases. 2008;**14**(2):147-161

[47] Bezirtzoglou E, Tsiotsias A, Welling GW. Microbiota profile in feces of breast-and formula-fed newborns by using fluorescence in situ hybridization (FISH). Anaerobe. 2011;**17**(6):478-482

[48] Ingber DE. Reverse engineering human pathophysiology with organs-on-chips. Cell. 2016;**164**(6):1105-1109

[49] Shankar V, Hamilton MJ, Khoruts A, Kilburn A, Unno T, Paliy O, et al. Species and genus level resolution analysis of gut microbiota in

Clostridium difficile patients following fecal microbiota transplantation. Microbiome. 2014;**2**(1):13

[50] Sanger F, Nicklen S, Coulson AR. DNA sequencing with chain-terminating inhibitors. Proceedings of the National Academy of Sciences. 1977;**74**(12):5463-5467

[51] Rhodes AN, Urbance JW, Youga H, Corlew-Newman H, Reddy CA, Klug MJ, et al. Identification of bacterial isolates obtained from intestinal contents associated with 12,000-year-old mastodon remains. Applied and Environmental Microbiology. 1998;**64**(2):651-658

[52] Wagner J, Coupland P, Browne HP, Lawley TD, Francis SC, Parkhill J. Evaluation of PacBio sequencing for full-length bacterial 16S rRNA gene classification. BMC Microbiology. 2016;**16**(1):274

[53] Loman NJ, Pallen MJ. Twenty years of bacterial genome sequencing. Nature Reviews Microbiology. 2015;**13**(12):787-794

[54] Petriz BA, Franco OL. Metaproteomics as a complementary approach to gut microbiota in health and disease. Frontiers in Chemistry. 2017;**5**:4

[55] Franzosa EA, Morgan XC, Segata N, Waldron L, Reyes J, Earl AM, et al. Relating the metatranscriptome and metagenome of the human gut. Proceedings of the National Academy of Sciences. 2014;**111**(22):E2329-E2338

[56] Gosalbes MJ, Durbán A, Pignatelli M, Abellan JJ, Jiménez-Hernández N, Pérez-Cobas AE, et al. Metatranscriptomic approach to analyze the functional human gut microbiota. PLoS One. 2011;**6**(3):e17447

[57] Xiong W, Abraham PE, Li Z, Pan C, Hettich RL. Microbial metaproteomics for characterizing the range of metabolic functions and activities of human gut microbiota. Proteomics. 2015;**15**(20):3424-3438

[58] Maier TV, Lucio M, Lee LH, VerBerkmoes NC, Brislawn CJ, Bernhardt J, et al. Impact of dietary resistant starch on the human gut microbiome, metaproteome, and metabolome. MBio. 2017;**8**(5):e01343-e01341

[59] Scholz M, Ward DV, Pasolli E, Tolio T, Zolfo M, Asnicar F, et al. Strain-level microbial epidemiology and population genomics from shotgun metagenomics. Nature Methods. 2016;**13**(5):435-438

[60] Leone V, Gibbons SM, Martinez K, Hutchison AL, Huang EY, Cham CM, et al. Effects of diurnal variation of gut microbes and high-fat feeding on host circadian clock function and metabolism. Cell Host & Microbe. 2015;**17**(5):681-689

[61] Roume H, Muller EE, Cordes T, Renaut J, Hiller K, Wilmes P. A biomolecular isolation framework for eco-systems biology. The ISME Journal. 2013;**7**(1):110

[62] Huan T, Forsberg EM, Rinehart D, Johnson CH, Ivanisevic J, Benton HP, et al. Systems biology guided by XCMS online metabolomics. Nature Methods. 2017;**14**(5):461-462

[63] Tarca AL, Carey VJ, Chen XW, Romero R, Drăghici S. Machine learning and its applications to biology. PLoS Computational Biology. 2007;**3**(6):e116

[64] Rahman SF, Olm MR, Morowitz MJ, Banfield JF. Machine learning leveraging genomes from metagenomes identifies influential antibiotic resistance genes in the infant gut microbiome. mSystems. 2018;**3**(1):e00123-e00117

[65] Ramachandran G, Bikard D. Editing the microbiome the CRISPR way. Philosophical Transactions of the Royal Society B. 2019;**374**(1772):20180103

[66] Burstein D, Harrington LB, Strutt SC, Probst AJ, Anantharaman K, Thomas BC, et al. New CRISPR—Cas systems from uncultivated microbes. Nature. 2017;**542**(7640):237-241

Section 2

Microbiota in Health and Diseases

Chapter 2

"Dialogue" between the Human Microbiome and the Brain

Natalia Beloborodova and Andrey Grechko

Abstract

In conditions of severe gut dysbiosis, there is a risk of developing diseases of the host organism in general and of the brain in particular, as evidenced by a growing number of studies. This chapter focuses on several groups of low-molecular-weight compounds that originate primarily from the gut microbiota. It discusses the results of experimental and clinical studies on the effect of microbial metabolites (such as short-chain fatty acids, phenolic metabolites of tyrosine, indolic metabolites of tryptophan, trimethylamines) on the brain. Several studies have proven that the microbial metabolite profiles in the gut and serum are interlinked and reflect a disruption of the gut microbial community. Using 16S ribosomal RNA gene sequencing, it was found that the gut microbiota of patients with positive or negative dynamics of neurological status differ taxonomically. The chapter also presents data obtained from animal germ-free (GF) models. Many researchers would like to consider the gut microbiota as a new therapeutic target, including for the treatment of brain diseases, stroke prevention, reduction of neuroinflammation, and more successful neurorehabilitation of patients.

Keywords: human microbiome, microbial metabolites, brain damage, gut microbiota dysbiosis, mental health, Alzheimer's disease, autism, stroke, critical ill patients, neurorehabilitation

1. Introduction

The human gut microbiome is a community of trillions of microorganisms that produce and use many molecules of microbial origin. Normally, the epithelial–immune–gut barrier supports homeostasis in the host body. The importance of the function of the gut microbiota for the host organism allows us to consider it as a large but "invisible organ" [1]. In conditions of severe gut dysbiosis, there is a risk of developing diseases of the host organism in general and of the brain in particular, as evidenced by a growing number of studies [2, 3]. The relevance of studying the relationship between the human microbiome and the brain is confirmed by a 20-fold increase in the number of publications on this topic in the PubMed database over the past 10 years (**Figure 1**).

Today, modern technologies allow us to identify hundreds of types of microorganisms in the human gut. Various microbial metabolites are also available for measurement in biological material samples, including feces, blood, urine, cerebrospinal fluid (CSF), and so on [1–3]. Thus, the possibilities of determining microbiota metabolites have expanded to studying their role both in healthy people and in patients with various diseases.

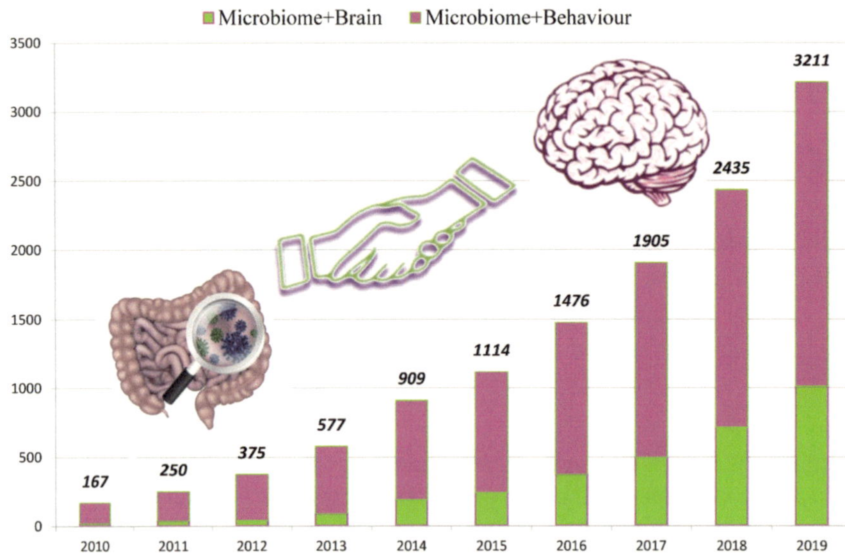

Figure 1.
Graph showing a 20-fold increase from 2010 to 2019 in the number of publications on the relationship between the human microbiome and the brain, according to PubMed. Keyword search results: microbiome and brain, microbiome and behaviour.

The results of numerous studies show that the gut microbiota affects the development of diseases of the central nervous system (CNS), including motor and behavioral disorders, neurodegenerative diseases, and cardiovascular and neuroimmune-mediated disorders [4–6]. The existence of the microbiome–gut–brain axis is now generally recognized. There are several different mechanisms of gut bacteria action on the nervous system, including changes in the activity of the stress-related hypothalamic–pituitary–adrenal axis, vagus nerve stimulation, and the secretion of short-chain fatty acids (SCFAs), which can activate microglial cells and affect the permeability of the blood–brain barrier. Evolutionarily conserved signals that are involved in the communication between microbiota and the host, which include different neuroactive substances, are known as neurochemicals [7].

This chapter focuses on several groups of low-molecular-weight compounds that originate primarily from the gut microbiota; their involvement in the interaction of the microbiota and the brain has been studied in various experimental and clinical studies.

2. Some molecules involved in the "dialogue"

2.1 Short-chain fatty acids

SCFAs as byproducts of microbiota fermentations are widely studied. It is proven that microbial SCFAs (acetate, propionate, butyrate) are involved in the energy metabolism of the host [8, 9]. Attempts to cope with metabolic disorders in several diseases, including those of the brain, with the help of diets were unsuccessful. One study found different amounts of SCFAs were produced in the guts of subjects following the same diet (in terms of the amount and composition of fiber), since initially different gut microbiota can trigger different fermentation pathways of indigestible carbohydrates [6].

In their review, Dalile et al. [10] describe the effects of SCFAs on cellular systems and their interaction with gut–brain signaling pathways through immune, endocrine, neural, and humoral mechanisms. The researchers concluded that SCFAs can penetrate the blood–brain barrier (BBB) to directly interact with brain tissues and even contribute to strengthening the integrity of the BBB. In addition, SCFAs promote serotonin biosynthesis and affect the levels of certain neurotrophic proteins, in particular, brain-derived neurotrophic factor (BDNF) and glial cell line-derived neurotrophic factor (GDNF) [10]. SCFAs also interfere with pathological mechanisms that are important for Alzheimer's disease. Thus, SCFAs are able to inhibit the formation of soluble beta-amyloid (Aβ) aggregates, which are associated with synaptic dysfunction and neurotoxicity. Another study examined the formation of neurotoxic amyloid aggregates (in vitro) and the dose-dependent effects of individual SCFAs on this process [11]. The authors call for the development of a new generation of probiotics that can metabolize individual dietary fibers to form valerian, butyric, and propionic acids and thus reduce the risk of developing neurodegenerative disorders. Unfortunately, animal and in vitro studies using pure fatty acid substances have several limitations. The source of SCFAs in vivo is the gut microbiota, and it remains unclear whether physiologically significant concentrations of SCFAs can be created in the human brain [10].

2.2 Metabolites of aromatic amino acids

Tyrosine and tryptophan are two of the nine essential amino acids that cannot be synthesized in the human body. Various metabolic pathways of metabolism of aromatic amino acids, such as tyrosine and tryptophan, with different endogenous and microbial enzymes, have been previously described [1]. Most often, the products of microbial protein biodegradation are associated with negative or toxic effects [2, 12]. At the same time, results of various studies suggest that the products of anaerobic bacteria from a healthy human gut (metabolites of some *Clostridium* species, *Bacteroids*, *Bifidobacterium*, etc.) can be useful [8, 11–13], including for brain function, which we discuss later in the chapter.

2.2.1 Phenolic metabolites of tyrosine

Phenylcarboxylic acids (PhCAs) are metabolites of tyrosine that circulate in the blood of a healthy human in a constantly low concentration, normally not exceeding 5 μM [12]. Their microbial origin has been proven [12, 13], as have the causes of a significant increase in the number of certain PhCAs, such as *p*-HPhLA, PhLA, and p-HPhAA, in the blood serum of patients with sepsis and sepsis-associated encephalopathy [2]. Serum and fecal profiles of these aromatic microbial metabolites reflect gut microbiota disruption in critically ill patients, including those with brain pathology. It has been shown that the aromatic microbial metabolite profiles in the gut and serum are interlinked and reflect a disruption of the gut microbial community [14].

The taxonomic composition of microbiota and the profile of microbial metabolites of PhCAs were studied in critically ill patients with severe brain damage in comparison with other groups of patients, including healthy individuals. Using the 16S-ribosomal RNA (16S-*r*RNA) gene sequencing method, it was found that patients with positive dynamics were more characterized by a shift in the balance of the gut microbiota towards the predominance of *Clostridium* taxa [14]. The Glasgow Coma Scale (GCS), the National Institutes of Health Stroke Scale (NIHSS), the Rivermead Mobility Index Scale, and the Rankin scale were used to assess neurological status over time, while the monitoring of serum PhCAs levels was performed by gas chromatography–mass spectrometry (GC–MS). Results showed that the positive dynamics of neurological status in patients with brain damage was associated with serum level of phenylpropionic acid (PhPA) [15]. Based on studies that have established that PhPA is the end product of tyrosine metabolism by *Clostridia sporogenes*

[16, 17], we believe that special attention should be paid to further confirmation of the involvement of *C. sporogenes* and studying the pathophysiological role of its metabolites in the process of neurorehabilitation.

2.2.2 Indolic metabolites of tryptophan

The essential amino acid tryptophan is the only amino acid that contains the structure of an indole-bicyclic compound consisting of a six-membered benzene ring connected to a five-membered N-containing pyrrole ring, according to the Human Metabolome Database. Tryptophan is absorbed in the small intestine and metabolized to kynurenine, serotonin, and melatonin via the host's endogenous pathways. Manipulating heavily depleted tryptophan by way of diet has helped to identify patients who are prone to depression or other mood-lowering symptoms associated with dysfunctional monoaminergic systems, which can be attributed to serotonin deficiency [18]. The part of tryptophan that reaches the colon can be catabolized by the gut bacteria resulting in a variety of indole derivatives, such as indole, tryptamine, indoleethanol, indolepropionic acid (IPA), indolelactic acid (ILA), indoleacetic acid (IAA), skatole, indolealdehyde (IAld), and indoleacrylic acid [18, 19]. It is known that some products of bacterial biodegradation of tryptophan can be toxic, for example, indole, as well as indoxyl sulfate (IS), which is produced in the liver from indole and has a cytotoxic effect in high concentrations [19]. However, research shows that microbial tryptophan metabolites may also have a positive impact on host physiology. Tryptophan metabolites can modulate both the function of intestinal immune cells and astrocytes in the CNS via the aryl hydrocarbon receptor (AHR) [19, 20]. In experimental autoimmune encephalomyelitis, the effect of limiting inflammation of the CNS by affecting astrocytes in mice treated with antibiotics was shown by adding microbial metabolites of tryptophan from the gut microbiota (indole, indoxyl-3-sulfate, IPA, IAld) or the bacterial enzyme tryptophanase as AHR agonists [20].

Several studies have noted that IPA and IAA have anti-oxidative and anti-inflammatory effects. A comparison of the varying data on the blood concentrations of IPA and IAA in patients with different diseases suggests that levels of both indole metabolites (IPA and IAA) are reduced in cancer [21]. Unfortunately, no studies to date have analyzed the behavior of these metabolites in patients with brain tumors, which could be extremely interesting.

There is information about the bacteria of the gut microbiota that is associated with the production of specific metabolites of indole. Interestingly, many species of anaerobes from different families are able to carry out biotransformation of tryptophan in vitro with the formation of IAA (nine species of *Clostridium*, four of *Bacteroides*, three of *Bifidobacterium*, and one of *Peptostreptococcus*). However, the ability to produce IPA was found only in three *species* of *Clostridiaceae*, and one of *Peptostreptococcus* [16, 21]. At the same time, the results obtained in vivo are more modest. In an experimental study of germ-free (GF) mice, production of IPA was completely dependent on gut colonization only by *C. sporogenes* [22].

The severity of stroke outcome in patients is associated with a stroke-induced inflammatory response, which in turn is linked with an increase in tryptophan catabolism [23, 24]. In Parkinson disease (PD) patients, CSF levels of tryptophan and kynurenic acid have been found to be significantly lower compared to healthy controls [25]. Future investigations are required to decipher how tryptophan metabolites derived from microbes are linked to inflammation in brain disorders [5]. The search and modification of methods for accurate measurement of microbial tryptophan metabolites continues. The availability of methods for determining concentrations of microbial tryptophan metabolites in serum and CSF is currently limited and better quantitative analytical methods targeting a larger variety of microbial tryptophan metabolites are needed [26].

2.3 Trimethylamines

The formation of trimethylamine (TMA) occurs in the intestine via biotransformation of dietary lecithin, choline, or L-carnitine found in certain animal products (red meat, egg yolks) and is associated with bacteria of the genera *Anaerococcus*, *Clostridium*, *Escherichia*, *Proteus*, *Providencia*, and *Edwardsiella*. It is known that TMA is absorbed into the blood and oxidized in the liver by the flavin monooxygenase enzyme to form trimethylamine N-oxide (TMAO) [27]. TMAO is found in CSF, indicating its ability to penetrate the blood–brain barrier [28].

The role of TMAO in neurodegenerative diseases, including AD, has been investigated extensively in the last five years. A study by Xu et al. [29] analyzed 20 metabolites that are significantly associated with cognitive decline in patients with AD. Potential genetic pathways underlying the strong association between TMAO and AD have been investigated. Employing an integrated computational approach, researchers identified nine main pathways and found that AD is closely related to TMAO. Thus, common genetic pathways underlying known biomarkers of AD were identified, with TMAO identified as the top-ranked microbial metabolite [29].

Researchers studied TMAO as a biomarker of AD by comparing three groups of patients: those with AD clinical syndrome, those with mild cognitive impairment (MCI), and cognitively unimpaired individuals. All patient groups had undergone lumbar puncture with CSF collection (n = 410), as well as TMAO and other biomarkers of AD quantification. Metabolites of microbiota TMAO were significantly elevated in CSF and associated with other biomarkers of AD pathology (phosphorylated tau and phosphorylated tau/Aβ42) and neuronal degeneration (total tau and neurofilament light chain protein), which confirms gut microbial involvement in AD [30].

2.4 Neurotransmitters

The gut microbiota can produce and/or consume numerous neurotransmitters, including dopamine, norepinephrine, serotonin, or gamma-aminobutyric acid (GABA) [4, 31]. Microbiota-dependent effects on gut serotonin significantly impact host physiology. For example, it is known that the gut contains the bulk of the body's serotonin (more than 85 percent 5-hydroxytryptamine (5-HT)), but the mechanisms that control the metabolism of 5-HT obtained from the gut are still unclear. A mammalian experiment showed that indigenous spore-forming bacteria from mouse and human microbiota promote 5-HT biosynthesis from colonic enterochromaffin cells, which supply 5-HT to the mucosa, lumen, and circulating platelets [32].

3. Special experimental models

Disorders of the gut microbiome have been experimentally documented in some brain diseases and stroke. In animal models of AD, PD, and acute stroke, dysbiosis, intestinal motility disorders, and/or increased intestinal permeability were demonstrated. A pro-inflammatory immune response and increased microglia reactivity were recorded, compared with a non-diseased condition. Special experimental models of non-microbial or GF animals were used to determine the influence of microbiota on the mechanisms of stroke development [33].

From these gnotobiotic animals, it is possible to decipher mechanisms of communication between specific members of the microbiota and the host organism. Animals lacking microbiota have extraordinarily different development and physiology than animals hosting commensal bacteria. GF animals have impaired immune systems, dysregulated hormone signaling, altered metabolism, and differences in neurotransmission from their conventional counterparts [34, 35].

GF mice show an underdeveloped microglia phenotype, which is manifested by an incomplete immune response to damage. In an experimental stroke model, GF mice showed an incomplete response to brain damage; there was no delineation of the damage locus, which was manifested by an increase in the volume of damage compared to normal animals. Thus it was determined that the microglia of GF animals is morphologically immature [36].

The most common form of dementia is AD, a neurodegenerative disorder associated with impaired cognitive function. This pathology is characterized by extracellular beta-amyloid (Aβ) plaques and intracellular neurofibrillary tangles composed of hyperphosphorylated tau protein [37].

When studying the connection of microbiota with the brain, one of the tasks is to find evidence of bacterial participation in AD pathogenesis through the formation of amyloid. The results of an experimental model of AD on transgenic mice revealed a tendency to the expression of amyloid precursor protein-β (APP). When these mice were kept in non-microbial conditions, cerebral β-amyloid plaques were less developed than in a normal environment [38]. This experiment indicates that the microbiota is involved in triggering adverse changes in the brains of transgenic animals, but undoubtedly this depends on the species composition and metabolic activity of the bacteria. For example, in AD participants, the gut microbiome has a reduced microbial diversity and taxonomically differs from the control age and sex correspondences of individuals, in particular, in AD compared to the control the number of *Firmicutes* and *Bifidobacterium* was reduced, but the number of *Bacteroidetes* was increased. The potential amyloidogenic properties of gut bacteria were evaluated and the composition of the microbiota and the aggregation of cerebral amyloid-β were also influenced by nutrients [11, 39, 40].

Several studies have reported that the microbiome of young mice differs significantly from that of older mice, in particular in the ratio of *Firmicutes* to *Bacteroidetes*. The benefits of the microbiota of young mice were demonstrated in an experiment on stroke models, in which transplantation of the gut microbiota from young to old mice contributed to an improvement in the outcome of stroke [41].

There are some limitations in experiments with GF animals because animals with a diverse microbiota have more developed intestinal epithelium than GF animals, which affects the functioning of the body as a whole. Studying the participation of microbiota in the functioning of the brain may be not always correct in case of comparison of the results obtained in GF and normal animals. The new approach avoids these difficulties by using special mice with a modified microbiota, called the altered Schaedler flora (ASF) mouse line, because they are colonized by only eight species of known bacteria [42].

The majority of research showing that microbiota can influence the nervous system has been performed in animals. As such, there is a strong need for well-designed human cohorts. Neuroactive compounds of microbial origin can directly modulate not only neuronal function and plasticity but also human behavior also [5].

4. Microbiome and human mental health

The metabolism of tryptophan via the kynurenine pathway leads to the formation of kynurenine and its neuroactive metabolites, such as 3-hydroxykynurenine, kynurenic acid, quinolic acid, and xanthurenic acid. The involvement of kynurenine and its metabolites in the pathogenesis of depressive disorders and schizophrenia is being studied [43]. For example, in patients with schizophrenia, an increased concentration of 3-hydroxykynurenine in the blood was measured. It is important to note that after targeted treatment, the level of this metabolite was normalized. This fact indirectly confirms the initial violation of tryptophan metabolism along the kynurenine pathway in schizophrenia [44].

According to the 2016 report, "The Five Year Forward View for Mental Health," from the independent Mental Health Taskforce to the NHS in England, mental disorders in the modern world affect every fourth person on the planet [45], which is a serious justification for the search for new mechanisms of the influence on mental status, including by studying and correcting the microbiome.

A clinical study examined how the gut microbiota and its associated metabolites were changed in sleep disorders in children with autism spectrum disorders (ASD). There was a decrease in the abundance of *Faecalibacterium* and *Agatobacterium*, a decrease in 3-hydroxybutyric acid and melatonin, and an increase in serotonin levels. These changes can worsen sleep problems and major symptoms in children with ASD [46].

Some studies have reported interesting correlations between severity of behavioral and gastrointestinal symptoms; others have demonstrated potential benefits of probiotics in correcting dysbiosis and reducing the severity of ASD symptoms. The general conclusion of these studies is that future research based on more randomized controlled studies with larger population sizes and standardized use of strains, concentration of probiotics, duration of treatments, and methods of DNA extraction is needed in this area, which may lead to more robust results [47].

According to the World Health Organization, mental disorders are quite common even in people who lead a seemingly normal lifestyle [48]. At the same time, new evidence suggests that less than 10 percent of mental and neurodegenerative diseases have a strict genetic etiology. Other predisposing and concomitant factors, such as stress, environmental exposures to potentially toxic elements, and other factors may influence neurometabolism, which may increase the risk for depression, autism, sclerosis, PD, and AD [49, 50]. Among these factors, an important place is occupied by the gut–brain microbiome relationship at the level of metabolomic connections, which allow us to conceptually rethink the causes and mechanisms of mental health disorders. Possibly in some categories of people with predisposition, the metabolic activity of the gut microbiome may affect not only the development, but also the severity of depressive disorder [51].

5. Microbiome and inflammatory events after stroke

The gut inflammatory and immune response can play a key role in the pathophysiology of severe course and development of complications after stroke. This can be judged by studying the mechanisms that occur in the brain when damaged. Proinflammatory T cells are often associated with increased inflammatory damage, but research of the gut inflammatory and immune response after stroke is still in its initial stage [52]. It would be crucial to understand which metabolites from the gut microbiome may affect the degree of brain damage, stroke outcome, and concomitant post-stroke diseases.

An experimental stroke model of GF mice clearly demonstrated the role of microbiota. When the mice were recolonized using a dysbiotic post-stroke microbiota, an increase in the volume of brain damage and functional deficit was observed [53]. In another experimental study, after the use of a cocktail of antibiotics in animals, a significant decrease in the volume of the heart attack in the acute phase of stroke was observed. The neuroprotective effect was varied depending on the type of antibiotic and correlated with the specific microbial population, rather than with the overall bacterial density. In particular, a link was found between the large and small size of a brain infarction and the enzymatic pathway of the aromatic metabolism in certain strains of *Bacteroidetes* [54].

In clinical pilot research, which included patients with severity of neurologic deficit, the taxonomic composition of the gut microbiota using real-time polymerase chain reaction (PCR) was studied. Correlation analysis revealed some connection between microbiology and clinical and laboratory indicators, for example, strong negative correlations between Glasgow coma scale scores and the abundance of *Enterococcus* spp. ($r = -0.77$, $p < 0.05$). It is interestin that statistically significant negative correlations between cortisol levels and the abundance of *B. thetaiotaomicron* or *F. prausnitzii* ($r = -0.57$, $r = -0.62$, respectively) were detected only in patients in a vegetative state [55].

Many authors report dysbiosis in stroke patients [56–58]. Some authors associate the dominance of SCFA producers, such as *Akkermansia*, *Odoribacter*, *Ruminococcaceae*, and *Victivallis*, with positive clinical outcomes, while the genus *Enterobacter* had significantly negative correlation with the dynamics of neurological status [56]. At the same time, in another study, *Akkermansia* was reduced in patients with cerebral infarction compared with a group of healthy people ($p < 0.05$) [57].

Pluta et al. [58] presented taxonomic findings in stroke patients. The authors launched an active discussion and tried to find explanations for the metabolic features of various genera and types of microbes, which, according to various data, dominated in the gut of stroke patients. For example, *A. muciniphila* uses mucin to produce acetate, which can be used by other bacteria, such as bacteria from the Ruminococcaceae and Odoribacter families, to produce butyrate [59]. However, despite many studies in this direction, significant differences and even sometimes contradictions of taxonomic findings lead us to conclude that the available information is not enough to form a coherent hypothesis.

It should be noted that the data on the taxonomic composition of the gut microbiota in most studies were obtained by examining samples from patients in the early stages (first and second day) after a stroke. The study of the composition of the gut microbiota in patients with a complicated course after stroke is even more relevant. These patients need intensive care for a long time due to the development of so-called chronic critical illness (CCI) [60]. Loss of microbial diversity and pathogen domination of the gut microbiota has been noted in such patients [61]. Significant differences were found for four genera: *Prevotella*, *Klebsiella*, *Streptococcus*, and *Clostridium XI* [62], which were previously mentioned in connection with some neuropsychiatric disorders [63, 64].

The interrelation of factors influencing the development of a CCI as a result of long-term violation of the functions of the brain and the gut microbiota has been studied [15]. The results confirm the association of taxonomic composition and profile of certain aromatic metabolites of the gut microbiota with the progression or reversibility of neurological disorders in CCI patients. A gross imbalance of microbial metabolism contributes to the formation of general metabolic dysfunction of the human body (**Figure 2**).

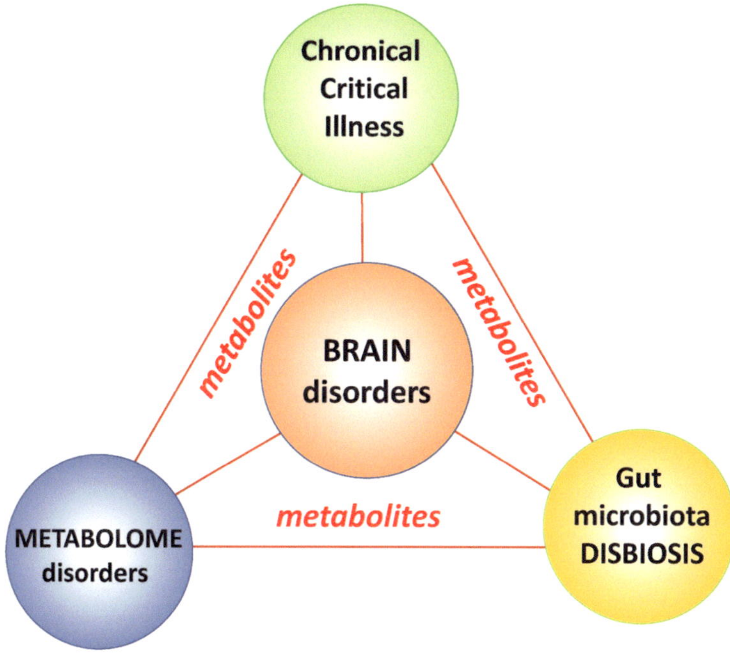

Figure 2.
Post-stroke complications and mechanisms of chronic critical illness are closely related to taxonomy disorders and metabolic dysfunction in the gut microbiota.

It is important to remember that microbial diversity and composition of the microbiota can be influenced by many personal and environmental factors (diet, infection, concomitant diseases, use of antibiotics and other medications, social stress, etc.), which can significantly affect the microbiota–gut–brain axis at all stages of life [65]. This fact should be considered in the future when developing methods to correct the dysfunction of the microbiota.

6. Conclusion

Due to growing interest in the human microbiome and rapid development of diagnostic technologies, the taxonomy of the gut microbiota in various diseases and disorders of the brain is quickly accumulating. Most researchers are coming to a common understanding of the importance of the communication between the human microbiome and the brain and are investigating binding small molecules as biomarkers and pathophysiological effects. Soon, the significance of particular microbial metabolites in the human metabolome will be evaluated in more detail. Figuratively speaking, this will allow us to master the "language" of the "dialogue" between the microbiome and the brain. Already, many researchers would like to consider the gut microbiota as a new therapeutic target, including for the treatment of brain diseases, stroke prevention, reduction of neuroinflammation and more successful neurorehabilitation of patients.

Author details

Natalia Beloborodova* and Andrey Grechko
Federal Research and Clinical Center of Intensive Care Medicine and Rehabilitology, Moscow, Russian Federation

*Address all correspondence to: nvbeloborodova@yandex.ru

References

[1] Beloborodova N.V., Grechko A.V., Olenin A.V. Metabolomic Discovery of Microbiota Dysfunction as the Cause of Pathology, Metabolomics - New Insights into Biology and Medicine, Wael N. Hozzein, Published: July 1st 2020 IntechOpen, DOI: 10.5772/intechopen. 87176.

[2] Beloborodova N.V., Olenin A.Yu., Pautova A.K. Metabolomic findings in sepsis as a damage of host-microbial metabolism integration. // JOURNAL OF CRITICAL CARE, 2018, 43, 246-255, https://doi.org/10.1016/j.jcrc.2017.09.014

[3] Benakis C, Martin-Gallausiaux C, Trezzi JP, Melton P, Liesz A, Wilmes P. The microbiome-gut-brain axis in acute and chronic brain diseases. Curr Opin Neurobiol. 2020;61:1-9. doi:10.1016/j. conb.2019.11.009

[4] Vogt NM, Kerby RL, Dill-McFarland KA, Harding SJ, Merluzzi AP, Johnson SC, Carlsson CM, Asthana S, Zetterberg H, Blennow K, Bendlin BB, Rey FE. Gut microbiome alterations in Alzheimer's disease. Sci Rep. 2017 Oct 19;7(1):13537. doi: 10.1038/ s41598-017-13601-y. PMID: 29051531; PMCID: PMC5648830

[5] Strandwitz P. Neurotransmitter modulation by the gut microbiota. Brain Res. 2018 August 15; 1693(Pt B): 128-133. doi:10.1016/j.brainres.2018.03.015

[6] Cani PD, Van Hul M, Lefort C, Depommier C, Rastelli M, Everard A. Microbial regulation of organismal energy homeostasis. Nat Metab. 2019;1(1):34-46. doi:10.1038/s42255-018-0017-4

[7] Oleskin AV, Shenderov BA. Probiotics and Psychobiotics: the Role of Microbial Neurochemicals. Probiotics Antimicrob Proteins. 2019 Dec;11(4):1071-1085. doi: 10.1007/s12602-019-09583-0. PMID: 31493127 Review.

[8] Oleskin AV, Shenderov BA. Neuromodulatory effects and targets of the SCFAs and gasotransmitters produced by the human symbiotic microbiota. Microb Ecol Health Dis. 2016;27:30971. Published 2016 Jul 5. doi:10.3402/mehd.v27.30971

[9] Schonfeld P., Wojtczak, L. Short- and medium-chain fatty acids in energy metabolism: the cellular perspective. J. Lipid Res. , 2016, 57, 943-954.

[10] Dalile B, Van Oudenhove L, Vervliet B, Verbeke K. The role of short-chain fatty acids in microbiota-gut-brain communication. Nat Rev Gastroenterol Hepatol. 2019;16(8):461-478. doi:10.1038/s41575-019-0157-3

[11] Ho L, Ono K, Tsuji M., Mazzola P., Singh,R., Pasinetti G. M. Protective roles of intestinal microbiota derived short chain fatty acids in Alzheimer's disease-type beta-amyloid neuropathological mechanisms. Expert Rev. Neurother. 2018, 18, 83-90. doi: 10.1080/14737175.2018.1400909

[12] Beloborodova NV, Khodakova AS, Bairamov IT, Olenin AY. Microbial origin of phenylcarboxylic acids in the human body. Biochemistry (Mosc). 2009 Dec;74(12):1350-1355. doi: 10.1134/ s0006297909120086. PMID: 19961416.

[13] Dodd D, Spitzer MH, Van Treuren W, et al. A gut bacterial pathway metabolizes aromatic amino acids into nine circulating metabolites. Nature. 2017;551(7682):648-652. doi:10.1038/nature24661

[14] Chernevskaya E, Beloborodova N, Klimenko N. et al. Serum and fecal profiles of aromatic microbial metabolites reflect gut microbiota disruption in critically ill patients: a prospective observational pilot study. Crit Care 24, 312 (2020). https://doi. org/10.1186/s13054-020-03031-0

[15] Chernevskaya E, Klimenko N, Pautova, A, Buyakova I, Tyakht A, Beloborodova N. Host-Microbiome Interactions Mediated by Phenolic Metabolites in Chronically Critically Ill Patients. Metabolites 2021, 11, 122. https://doi.org/10.3390/metabo11020122

[16] Elsden S R, Hilton M G, Waller J M. The end products of the metabolism of aromatic amino acids by clostridia. Arch. Microbiol.,1976, 107, 283-288

[17] Beloborodova NV, Bairamov IT, Olenin AY, Khabib ON, Fedotcheva NI. Anaerobic Microorganisms from Human Microbiota Produce Species-Specific Exometabolites Important in Heath and Disease. 2014. Global Journal of Pathology and Microbiology. Vol. 2. № 1. 43-53

[18] Palego L, Betti L, Rossi A, Giannaccini G. Tryptophan biochemistry: structural, nutritional, metabolic, and medical aspects in humans. J Amino Acids. 2016;2016:8952520. doi: 10.1155/2016/8952520.

[19] Roager HM, Licht TR: Microbial tryptophan catabolites in health and disease. Nat Commun 2018, 9:1-10 http://refhub.elsevier.com/S0959-4388(19)30108-4/sbref0275

[20] Rothhammer V, Mascanfroni ID, Bunse L, Takenaka MC, Kenison JE, Mayo L, Chao CC, Patel B, Yan R, Blain M, Alvarez JI, Kébir H, Anandasabapathy N, Izquierdo G, Jung S, Obholzer N, Pochet N, Clish CB, Prinz M, Prat A, Antel J, Quintana FJ. Type I interferons and microbial metabolites of tryptophan modulate astrocyte activity and central nervous system inflammation via the aryl hydrocarbon receptor. Nat Med. 2016 Jun;22(6):586-597. doi: 10.1038/nm.4106.

[21] Beloborodova NV, Chernevskaya EA, Getsina ML. Indolic Structure Metabolites as Potential Biomarkers of Non-infectious Diseases. Curr Pharm Des. 2021;27(2):238-249. doi: 10.2174/1381612826666201022121653.

[22] Wikoff WR, Anfora AT, Liu J, Schultz PG, Lesley SA, Peters EC et al. Metabolomics analysis reveals large effects of gut microflora on mammalian blood metabolites. PNAS 2009 March 10;106(10): 3698-3703 DOI:10.1073/pnas.0812874106

[23] Brouns R, Verkerk R, Aerts T, De Surgeloose D, Wauters A, Scharpe´ S, De Deyn PP: The role of tryptophan catabolism along the kynurenine pathway in acute ischemic stroke. Neurochem Res 2010, 35:1315-1322. http://refhub.elsevier.com/S0959-4388(19)30108-4/sbref0280

[24] Ormstad H, Verkerk R, Aass HCD, Amthor K-F, Sandvik L: Inflammation-induced catabolism of tryptophan and tyrosine in acute ischemic stroke. J Mol Neurosci 2013, 51:893-902. http://refhub.elsevier.com/S0959-4388(19)30108-4/sbref0285

[25] Szabo N, Kincses ZT, Toldi J, Ve´csei L: Altered tryptophan metabolism in Parkinson's disease: a possible novel therapeutic approach. J Neurol Sci 2011, 310:256-260. http://refhub.elsevier.com/S0959-4388(19)30108-4/sbref0290

[26] Pautova A, Khesina Z, Getsina M, Sobolev P, Revelsky A, Beloborodova N. Determination of tryptophan metabolites in serum and cerebrospinal fluid samples using microextraction by packed sorbent, silylation and GC-MS detection // Molecules, 2020, 25 (14), #3258 https://doi.org/10.3390/molecules25143258

[27] Romano KA, Vivas EI, Amador-Noguez D, Rey FE. Intestinal microbiota composition modulates choline bioavailability from diet and accumulation of the proatherogenic metabolite trimethylamine-N-oxide. MBio. 2015;6:e02481. doi:10.1128/mBio.02481-14.

[28] Del Rio D, Zimetti F, Caffarra P, Tassotti M, Bernini F, Brighenti F, et al. The gut microbial metabolite trimethylamine-N-oxide is present in human cerebrospinal fluid. Nutrients. 2017;9:1-7. https://doi.org/10.3390/nu9101053.

[29] Xu R, Wang Q. Towards understanding brain-gut-microbiome connections in Alzheimer's disease. BMC Syst Biol. 2016;10(Suppl 3):63. doi: 10.1186/s12918-016-0307-y

[30] Vogt NM, Romano KA, Darst BF et al. The gut microbiota-derived metabolite trimethylamine N-oxide is elevated in Alzheimer's disease. Alz Res Therapy 10, 124 (2018). https://doi.org/10.1186/s13195-018-0451-2

[31] Jameson KG, Hsiao EY. Linking the Gut Microbiota to a Brain Neurotransmitter. Trends Neurosci. 2018 Jul;41(7):413-414. doi: 10.1016/j.tins.2018.04.001.

[32] Yano JM, Yu K, Donaldson GP, Shastri GG, Ann P, Ma L, Nagler CR, Ismagilov RF, Mazmanian SK, Hsiao EY. Indigenous bacteria from the gut microbiota regulate host serotonin biosynthesis. Cell. 2015 Apr 9;161(2):264-276. doi: 10.1016/j.cell.2015.02.047. Erratum in: Cell. 2015 Sep 24;163:258.

[33] Luczynski P, McVey Neufeld KA, Oriach CS, Clarke G, Dinan TG, Cryan JF. Growing up in a Bubble: Using Germ-Free Animals to Assess the Influence of the Gut Microbiota on Brain and Behavior. Int J Neuropsychopharmacol 19: pyw020, 2016. doi:10.1093/ijnp/pyw020.

[34] Kawase T, Nagasawa M, Ikeda H, Yasuo S, Koga Y, Furuse M. Gut microbiota of mice putatively modifies amino acid metabolism in the host brain. Br J Nutr 2017, 117: 775-783,. doi:10.1017/S0007114517000678

[35] Neufeld KM, Kang N, Bienenstock J, Foster JA. Reduced anxiety-like behavior and central neurochemical change in germ-free mice. Neurogastroenterol Motil 23: 255-264.e119, 2011. doi:10.1111/j.1365-2982.2010.01620.x.

[36] Erny D, Hrab_e de Angelis AL, Jaitin D, Wieghofer P, Staszewski O, David E, Keren-Shaul H, Mahlakoiv T, Jakobshagen K, Buch T et al.: Host microbiota constantly control maturation and function of microglia in the CNS. Nat Neurosci 2015, 18:965-977

[37] Scheltens P, Blennow K, Breteler MMB, De Strooper B, Frisoni GB, Salloway S, et al. Alzheimer's disease. Lancet. 2016;388:505-517.

[38] Harach T, Marungruang N, Dutilleul N, Cheatham V, Mc Coy K, Neher J, Jucker M, Fåk F, Bolmont T. Reduction of Alzheimer's disease beta-amyloid pathology in the absence of gut microbiota. arXiv:1509.02273v2. 2015.]

[39] Jiang C, Li G, Huang P, Liu Z, Zhao B. The Gut Microbiota and Alzheimer's Disease. J Alzheimers Dis. 2017;58(1):1-15. doi: 10.3233/JAD-161141. PMID: 28372330

[40] Pistollato F, Sumalla Cano S, Elio I, Masias Vergara M, Giampieri F, Battino M. Role of gut microbiota and nutrients in amyloid formation and pathogenesis of Alzheimer disease. Nutr Rev. 2016 Oct;74(10):624-634. doi: 10.1093/nutrit/nuw023

[41] Spychala MS, Venna VR, Jandzinski M, Doran SJ, Durgan DJ, Ganesh BP, Ajami NJ, Putluri N, Graf J, Bryan RM et al.: Age-related changes in the gut microbiota influence systemic inflammation and stroke outcome. Ann Neurol 2018, 84:23-36.

[42] Lyte JM, Proctor A, Phillips GJ, Lyte M, Wannemuehler M. Altered Schaedler flora mice: a defined microbiota animal model to study the microbiota-gut-brain axis. Behav Brain Res 356: 221-226, 2019. doi:10.1016/j.bbr.2018.08.022.

[43] Shilov YE, Bezrukov MV. Kynurenines in pathogenesis of endogenous psychiatric disorders. Annals of the Russian Academy of Medical Sciences. 2013;68(1):35-41. (In Russ).] doi: 10.15690/vramn.v68i1.535].

[44] Myint AM, Schwarz M, Verkerk R. Imballance of kynurenine metabolites in drug naive schizophrenia. Brain Behav. Immun. 2011; 25 (8): 1576-1581.

[45] Mental Health Taskforce for NHS England The Five Year Forward View for Mental Health: A Report from the Independent Mental Health Taskforce to the NHS in England. London; February 2016

[46] Hua X, Zhu J, Yang T, Guo M, Li Q, Chen J, Li T. The Gut Microbiota and Associated Metabolites Are Altered in Sleep Disorder of Children With Autism Spectrum Disorders. Front. Psychiatry, 02 September 2020 https://doi.org/10.3389/fpsyt.2020.00855

[47] Abdellatif B, McVeigh C, Bendriss G, Chaari A. The Promising Role of Probiotics in Managing the Altered Gut in Autism Spectrum Disorders. Int J Mol Sci. 2020;21(11):4159. Published 2020 Jun 10. doi:10.3390/ijms21114159

[48] World Health Organization, Available from: https://www.who.int/news-room/fact-sheets/detail/depression Accessed 01.03.2021.

[49] Cabral Pinto MMS, Marinho-Reis AP, Almeida A, Ordens CM, Silva MMVG, Freitas S, Simões MR, Moreira PI, Dinis PA, Diniz ML, Ferreira da Silva EA, Condesso de Melo MT. Human predisposition to cognitive impairment and its relation with environmental exposure to potentially toxic elements. Environ Geochem Health. 2018 Oct;40(5):1767-1784. doi: 10.1007/s10653-017-9928-3.

[50] Levey DF, Polimanti R, Cheng Z. et al. Genetic associations with suicide attempt severity and genetic overlap with major depression. Transl Psychiatry 9, 22 (2019). https://doi.org/10.1038/s41398-018-0340-2

[51] Chen JJ, Zhou CJ, Zheng P, et al. Differential urinary metabolites related with the severity of major depressive disorder. Behav Brain Res. 2017;332:280-287. doi: 10.1016/j.bbr.2017.06.012

[52] Arya AK, Hu B. Brain-gut axis after stroke. Brain Circ. 2018;4(4):165-173. doi:10.4103/bc.bc_32_18

[53] Singh V, Roth S, Llovera G, et al. Microbiota Dysbiosis Controls the Neuroinflammatory Response after Stroke. J Neurosci. 2016;36(28):7428-7440. doi:10.1523/JNEUROSCI.1114-16.2016

[54] Benakis C, Poon C, Lane D, Brea D, Sita G, Moore J, Murphy M, Racchumi G, Iadecola C, Anrather J. Distinct Commensal Bacterial Signature in the Gut Is Associated With Acute and Long-Term Protection From Ischemic Stroke. Stroke. 2020 Jun;51(6):1844-1854. doi: 10.1161/STROKEAHA.120.029262. Epub 2020 May 14. PMID: 32404038

[55] Chernevskaya EA, Meglei AYu, Buyakova IV, Kovaleva NYu, Gorshkov KM, Zakharchenko VE, Beloborodova N.V. Taxonomic dysbiosis of gut microbiota and serum biomarkers reflect severity of central nervous system injury. Bulletin of RSMU. 2020; (5): –. DOI: 10.24075/brsmu.2020.053

[56] Li N, Wang X, Sun C, Wu X, Lu M, Si Y, Ye X, Wang T, Yu X, Zhao X. et al. Change of intestinal microbiota in cerebral ischemic stroke patients. BMC Microbiol. 2019, 19, 191.

[57] Ji W, Zhu Y, Kan P, Cai Y, Wang Z, Wu Z, Yang P. Analysis of intestinal microbial communities of cerebral infarction and ischemia patients based on high throughput sequencing technology and glucose and lipid

metabolism. Mol Med Rep. 2017 Oct;16(4):5413-5417. doi: 10.3892/mmr.2017.7227.

[58] Zeng, X.; Gao, X.; Peng, Y.; Wu, Q.; Zhu, J.; Tan, C.; Xia, G.; You, C.; Xu, R.; Pan, S.; et al. Higher risk of stroke is correlated with increased opportunistic pathogen load and reduced levels of butyrate-producing bacteria in the gut. Front. Cell. Infect. Microbiol. 2019, 9, 4.

[59] Pluta, R.; Januszewski, S.; Czuczwar, S.J. The Role of Gut Microbiota in an Ischemic Stroke. Int. J. Mol. Sci. 2021, 22, 915. https://doi.org/10.3390/ijms22020915

[60] Nelson, J.E.; Cox, C.E.; Hope, A.A.; Carson, S.S. Chronic critical illness. Am. J. Respir. Crit. Care Med. 2010, 182, 446-454.

[61] Ravi, A.; Halstead, F.D.; Bamford, A.; Casey, A.; Thomson, N.M.; Schaik, W.; Snelson, C.; Goulden, R.; Foster-Nyarko, E.; Savva, G.M.; et al. Loss of microbial diversity and pathogen domination of the gut microbiota in critically ill patients. Microb. Genom. 2019, 5, e000293.

[62] Zaborin A, Smith D, Garfield K, Quensen J, Shakhsheer B, Kade M, Tirrell M, Tiedje J, Gilbert J.A, Zaborina O. Membership and behavior of ultra-low-diversity pathogen communities present in the gut of humans during prolonged critical illness. mBio 2014, 5, e01361–e01314.

[63] Leslie DL, Kozma L, Martin A, Landeros A, Katsovich L, King RA, Leckman JF. Neuropsychiatric disorders associated with streptococcal infection: A case-control study among privately insured children. J. Am. Acad. Child. Adolesc. Psychiatry 2008, 47, 1166-1172.

[64] Lin P, Ding B, Feng C, Yin S, Zhang T, Qi X, Lv H, Guo X, Dong K, Zhu Y, Li Q. Prevotella and Klebsiella proportions in fecal microbial communities are potential characteristic parameters for patients with major depressive disorder. J. Affect. Disord. 2017, 207, 300-304.

[65] Cryan JF, O'Riordan KJ, Cowan CSM, Sandhu KV, Bastiaanssen TFS, Boehme M, Codagnone MG, Cussotto S, Fu¨ lling C, Golubeva AV et al.: The microbiota-gut-brain axis. Physiol Rev 2019, 99:1877-2013

Chapter 3

Intestinal Dysbiosis and Non-Alcoholic Fatty Liver Disease

Teresa Auguet, Laia Bertran and Jessica Binetti

Abstract

Non-alcoholic fatty liver disease (NAFLD) affects 20–30% of the population, with an increased prevalence in industrialized regions. Some patients with NAFLD develop an inflammatory condition termed non-alcoholic steatohepatitis (NASH) that is characterized by hepatocellular injury, innate immune cell-mediated inflammation, and progressive liver fibrosis. In clinical practice, abdominal imaging, which reveals hepatic steatosis, is sufficient for NAFLD diagnosis if other diseases have been rejected. However, a liver biopsy is needed to differentiate NASH from simple steatosis. Therapeutic strategies used to treat obesity and metabolic syndrome improve NAFLD, but there is no specific treatment effective for NASH. The gut microbiota (GM) is composed of millions of microorganisms. Changes in the GM have a significant impact on host health. Intestinal dysbiosis is an imbalance in the GM that can induce increased permeability of the epithelial barrier, with migration of GM-derived mediators through portal vein to the liver. These mediators, such as lipopolysaccharides, short-chain fatty acids, bile acids (BAs), choline, and endogenous ethanol, seem to be involved in NAFLD pathogenesis. Given this evidence, it would be interesting to consider GM-derived mediator determination through omics techniques as a noninvasive diagnostic tool for NASH and to focus research on microbiota modulation as a possible treatment for NASH.

Keywords: non-alcoholic fatty liver disease, non-alcoholic steatohepatitis, gut microbiota, intestinal dysbiosis, gut microbiota-derived mediators, noninvasive biomarker, therapeutic target

1. Introduction

Non-alcoholic fatty liver disease (NAFLD) is currently the most prevalent chronic liver disease worldwide [1]. A subset of NAFLD patients have the progressive form of NAFLD termed non-alcoholic steatohepatitis (NASH). NASH is typically characterized by a specific pattern on liver histology, including steatosis, lobular inflammation, and ballooning with or without perisinusoidal fibrosis [2]. It can progress to advanced fibrosis, cirrhosis, hepatocellular carcinoma, and liver-related morbidity and mortality. Liver disease is only the third leading cause of death in patients with NAFLD, following cardiovascular disease and malignancy [3].

Precise histological diagnosis of NAFLD is commonly based on liver biopsy [4]; however, biopsies present several potential problems [5]. Thus, there is a need for reliable and cost-effective noninvasive biomarkers to avoid the invasiveness of biopsy [6].

Although there are some clinical strategies to ameliorate NAFLD progression, such as treatments for obesity or type 2 diabetes mellitus (T2DM), there is no medication proven to be effective as a treatment for NASH [7]. Therefore, it is necessary to improve the research on possible therapeutic targets for NASH due to the severity of this pathological condition.

Previous evidences have linked gut dysbiosis with obesity, insulin resistance (IR), metabolic syndrome (MS), and NAFLD [8, 9]. The impact of the GM on NAFLD/NASH has been attributed to increased gut permeability, intestinal endotoxemia, endogenous alcohol production, upregulation of hepatic de novo lipogenesis and triglyceride synthesis, reduction in choline metabolism, and aggravation of IR [10]. The increased permeability of the intestinal barrier results in the release of substances such as lipopolysaccharides (LPS), bacterial components, short-chain fatty acids (SCFAs), bile acids (BAs), choline metabolites, and endogenous ethanol that reach the liver and seem to contribute to the pathogenesis of NAFLD (**Figure 1**) [11, 12]. It is important to note that some of these substances could perhaps be employed as potential noninvasive biomarkers of NAFLD progression.

Manipulation of the microbiota through probiotics, prebiotics, and antibiotic treatment yields encouraging results for the treatment of obesity, T2DM, and NASH in animal models, but data in humans are scarce. In regard to NAFLD, this

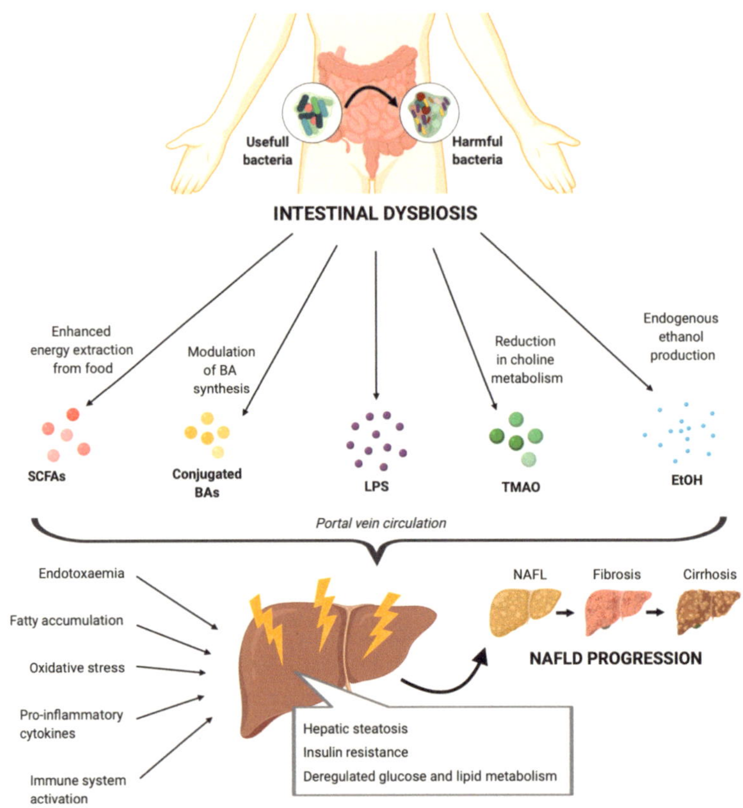

Figure 1.
Implication of intestinal dysbiosis in NAFLD pathogenesis. Short-chain fatty acids (SCFAs), bile acids (BAs), lipopolysaccharides (LPS), trimethylamine-N-oxide (TMAO), ethanol (EtOH), non-alcoholic fatty liver (NAFL), and non-alcoholic fatty liver disease (NAFLD).

therapeutic strategy seeks to prevent the endotoxicity produced by the microbiota-derived metabolites that reach the liver and promote the progression of the disease [13]. Thus, there is a need to focus research on the GM as a therapeutic target to ameliorate NASH.

To provide a broad overview of the relationship between intestinal dysbiosis and NAFLD, we have elaborated on this subject in this book chapter. In this sense, this narrative chapter will explain (a) non-alcoholic fatty liver disease, (b) the gut microbiota, (c) gut microbiota-derived mediators involved in NAFLD, and (d) the gut microbiota as a therapeutic target in NAFLD.

2. Non-alcoholic fatty liver disease

NAFLD has emerged as the most common form of chronic liver disease worldwide. The incidence of NAFLD has drastically increased in parallel with obesity in recent years. Currently, the global prevalence of NAFLD is approximately 25% [1], but it can increase to 58% in individuals who are overweight or as high as 98% in individuals with nondiabetic morbid obesity [14].

NAFLD comprises a spectrum of disorders extending from simple steatosis (SS) to NASH, fibrosis, and cirrhosis [2, 15]. This pathology has potentially serious sequelae [16]. Although SS tends to develop into a favorable clinical course [3], NASH can develop into liver cirrhosis and hepatocellular carcinoma [15]. Thus, liver-related mortality increases exponentially with an advance in the fibrosis stage [17]. In this regard, NASH is a very common cause of liver transplant worldwide [1]. Although the most common cause of death in patients with NAFLD is cardiovascular disease, independent of other metabolic comorbidities, NAFLD is becoming a major cause of liver disease-related morbidity (e.g., cirrhosis, end-stage liver disease, hepatocellular carcinoma, and liver transplantation).

NAFLD is characterized by significant lipid deposition in the hepatocytes of the liver parenchyma [18]. Obesity, T2DM, dyslipidemia, MS, and IR are the main risk factors for NAFLD [19]. Most NAFLD patients are asymptomatic, and the evidence of hepatic steatosis should be detected via a routine blood test, showing a deregulation in liver enzymes. Currently, it is not possible to diagnose NAFLD with only a blood test, but the aspartate aminotransferase (AST)-alanine aminotransferase ratio (ALT) can be used as a first step [20–22]. However, the ALT level correlation with histological findings has poor sensitivity and specificity for the diagnosis of NASH [23]. Then, it is necessary to rule out other causes of liver damage, such as alcoholic fatty liver disease, drug-induced liver injury, viral hepatitis, autoimmune liver disease, hemochromatosis, celiac disease, and Wilson's disease [1]. Finally, ultrasonography is the most common noninvasive tool used to detect NAFLD. There are also other imaging techniques used to detect liver steatosis, such as computer tomography or magnetic resonance imaging, but ultrasound is the technique that provides the most information without irradiation [24, 25].

One-third of the NAFLD-affected subjects progress to NASH. This condition is characterized by the presence of hepatocellular ballooning and inflammation and has a prevalence of 2–3% worldwide [2]. Key issues in NAFLD patients are the differentiation of NASH from SS and the identification of advanced hepatic fibrosis. To date, liver biopsy has been the *gold standard* for identifying these two critical end points but has well-known limitations, including invasiveness; rare but potentially life-threatening complications; poor tolerance; sampling variability; and cost. Furthermore, due to the epidemic proportion of individuals with NAFLD worldwide, liver biopsy evaluation is impractical, and noninvasive assessment for the diagnosis of NASH and fibrosis is needed [5]. NASH is confirmed when the hepatic tissue shows the presence of perilobular inflammation, hepatocellular ballooning, Mallory's hyaline, and acidophil bodies with or without fibrosis. Although there are other noninvasive tests, such as the fatty liver index, NAFLD fibrosis score, and FibroMeter, and elastographic techniques, such as FibroScan, that can suggest the presence of NASH and detect fibrosis [15], a precise histological diagnosis of NASH is commonly based on liver biopsy [26]. The development of alternative noninvasive strategies has been an area of intensive research over the past decade and currently.

Regarding NAFLD therapeutics, all forms of treatment of metabolic disorders are able to modify liver damage. Diet and lifestyle modification and insulin-sensitizing agents appear to be promisingly effective against NAFLD progression. However, these approaches may not be effective in some patients. Many other drugs are currently being studied to establish treatments for NAFLD. At present, no accepted drug treatment for NASH has been stated [24]. In this sense, it is very important to improve the knowledge of NAFLD physiopathology. Actually, the underlying precise mechanisms of NAFLD pathogenesis have just begun to be understood. The classic "multiple hit" theory states that lipid accumulation initiates hepatic steatosis and subsequently triggers multiple insults acting together (hormones/adipokines from adipose tissue, inflammation, deregulated fat metabolism, lipotoxicity, oxidative stress, mitochondrial dysfunction, and genetic and epigenetic factors), ultimately inducing NASH and cirrhosis [27]. Progression to NASH is linked to systemic inflammation, and it is associated with other pathological processes, such as innate immunity alterations, endoplasmic-reticulum stress, toll-like receptor (TLR) signaling, mitochondrial dysfunction, and intestinal dysbiosis [6, 28–32]. Regarding this last process, approximately 70–75% of blood that reaches the liver comes from the portal vein circulation that communicates the liver with the intestine [33]. The liver is continually exposed to GM-derived mediators, including bacteria and bacterial components, such as LPS, promoting an inflammatory response that contributes to liver injury [13].

3. The gut microbiota

Millions of symbiotic microorganisms live on and within human beings and play an important role in human health and disease. Initial colonization occurs at the time of birth, and humans progressively acquire $\sim 10^{14}$ bacterial cells at equilibrium, which remain for life [13].

The human microbiota, especially the GM, has even been considered to be an "essential organ," carrying approximately 150 times more genes than the human genome [34]. The GM is composed of an immense number of microorganisms (bacteria, viruses, and fungi) with several functions, such as host nutrition, bone mineralization, immune system regulation, xenobiotic metabolism, proliferation of intestinal cells, and protection against pathogens [35, 36]. This bacterial community is dominated by anaerobic bacteria and includes 500–1000 species [37]. *Firmicutes* and *Bacteroidetes* are the most important phyla among the intestinal bacteria, with a proportion of over 90% of the total community [38].

The duodenum and proximal jejunum normally contain small numbers of bacteria, usually lactobacilli and enterococci, which are facultative anaerobes. The distal ileum is a transition zone between sparse populations of aerobic bacteria of the proximal small intestine and very dense populations of anaerobic microorganisms in the large bowel. Occasional groups of bacteria can be found in low concentrations within the lumen of the small intestine. Bacteria do not form clusters, and the luminal contents are separated from the mucosa by a mucus layer [13].

The GM is specific to an individual and highly resilient to changes. However, it can be affected by several factors, intrinsic and extrinsic to the host, such as the subject's genetic makeup, dietary habits, antibiotic use, and environmental changes [13, 39, 40]. A disruption in the composition of the normal GM is known as intestinal dysbiosis [41, 42]. Generally, this process includes an unfavorable change in the bacterial composition, with a reduction in autochthonous bacteria and growth of others that prejudice host health [43].

3.1 Intestinal dysbiosis

Intestinal dysbiosis is a process that may adversely impact metabolism and produce immune responses, favoring NAFLD progression. Important studies on the relationship of the GM with obesity have identified profound changes in the composition and metabolic function of the GM in subjects with obesity. Moreover, these studies demonstrated that the GM interacts with host epithelial cells to indirectly control energy expenditure and storage and activate inflammatory responses in NASH pathogenesis [44]. Qualitative or quantitative imbalances in the GM might have serious health consequences for the host, including small intestinal bacterial overgrowth (SIBO) syndrome [13]. Due to gut dysbiosis, there is an elevated production of toxic bacterial components and metabolic mediators, which consequently accumulate in the intestine. In addition, an increase in intestinal permeability and further disruption of the epithelial barrier lead to the release of these GM-derived mediators [42], which could reach the liver through portal circulation, favoring hepatic inflammation and the development of NAFLD [45, 46]. After disruption of the gut epithelial barrier, the liver is exposed to microbial products and metabolites resulting from bacterial metabolism [47, 48]. In this sense, it has been demonstrated that patients with NAFLD have gut dysbiosis, gut epithelial barrier dysfunction, and increased translocation of bacterial components to the liver [49]. For this reason, mediators derived from gut dysbiosis might also be related to the pathogenesis of the disease. Several previous studies in clinical settings have associated intestinal dysbiosis with the occurrence of NAFLD [50–52] and with the progression to NASH [10, 53].

Among the various factors, dietary habits are considered to be most influential on the gut microbiome in subjects with obesity and NAFLD patients. It is well-known that a high-fat diet causes gut dysbiosis characterized by lowered species richness and changes in microbial composition, such as decreased *Bacteroidetes* and increased *Firmicutes* and *Proteobacteria* abundances [43]. On the other hand, *Prevotella*, a member of the phylum *Bacteroidetes*, is associated with plant-rich diets. *Prevotella*-dominated microbiotas have higher fiber utilizing capacity than *Bacteroides*-dominated microbiotas, producing higher amounts of SCFAs [54]. There are some studies that consider *Prevotella* to be a beneficial commensal bacterium [10, 55], but there are others that noted enriched fecal *Prevotella* in NASH or cirrhotic patients [56–58]. These contradictory results may be partly explained by the differences in populations, age, or NAFLD stages between the studies. In this sense, further studies on *Prevotella* should be directed to characterize properties at the species level and to evaluate these species in different stages of NAFLD.

GM-derived mediators resulting from intestinal dysbiosis could play a key role in NAFLD progression through several mechanisms: (1) enhanced energy extraction from food nutrients by formation of SCFAs; (2) modulation of BA synthesis, which is crucial for fat absorption and affects metabolism of glucose via farnesoid X receptor (FXR); (3) innate immune system activation by bacterial component translocation; (4) endogenous ethanol production; and (5) reduction in choline metabolism, which reduces efflux of very-low-density lipoprotein (VLDL) from hepatocytes, promoting inflammation. These mechanisms involve translocation of these mediators, such as SCFAs, BAs, endogenous ethanol, and choline metabolites, which may be potentially evaluated as noninvasive blood markers of NAFLD progression [59].

4. Gut microbiota-derived mediators involved in NAFLD

4.1 Short-chain fatty acids

SCFAs are molecules with seven carbon atoms or less, for example, acetic, propionic, and butyric acids, that are produced by the gut bacterial fermentation of cellulose, xylans, resistant starch, or inulin since humans lack enzymes that digest fibers. These substances can strongly regulate host metabolism [60]. In general, these SCFAs have several effects on energy metabolism, the immune response, and adipose tissue expansion and act as signaling molecules between the GM and the host. SCFAs provide not only important sources of nutrients and energy for the intestinal epithelium but also serve as precursors for lipogenesis and gluconeogenesis [61, 62]. SCFAs can directly act as lipid precursors in the liver and mediate other effects as ligands for G protein-coupled receptors, specifically the subtypes GPR41 and GPR43 [59]. Experimental studies have demonstrated that these SCFAs can modulate regulatory T-cell expansion and enhance neutrophil chemotaxis, promoting inflammation in mouse models [63–66]. Furthermore, SCFAs modulate the production of several inflammatory cytokines, including tumor necrosis factor (TNF)-α, interleukin 2 (IL-2), interleukin 6 (IL-6), and interleukin 10 (IL-10) [67]. Recently, some studies found that high concentrations of intestinal SCFAs as a result of dysbiosis and their G-protein coupled receptors play an important role in NAFLD progression [68, 69]. Activation of GPR41 and GPR43 stimulates secretion of peptide-YY, inhibits gut motility, and slows intestinal transit. Therefore, nutrient absorption and energy capture from the diet increase and may promote hepatic lipogenesis [56, 70]. Additionally, activation of GPR41 and GPR43 induces secretion of glucagon-like peptide-1 (GLP-1), which activates genes in hepatocytes that regulate fatty acid β-oxidation and insulin sensitivity [56, 71], promoting NAFLD occurrence and progression. Furthermore, clinical studies have demonstrated SCFA enrichment in fecal samples of children and adults with NAFLD [72, 73].

However, other previously published studies have reported that SCFAs could be beneficial in the progression of NAFLD. In this regard, butyrate activates AMP-activated protein kinase (AMPK) in the liver and accelerates the assembly of tight junction proteins [74, 75], improving intestinal barrier dysfunction and reducing metabolic endotoxemia. In addition, butyrate is able to modulate regulatory T-cell activity, suppressing the immune response and reducing liver inflammation [76].

The close relationship between intestinal dysbiosis and SCFA production, according to the results of previous experimental and clinical studies, provides evidence of their potential use as markers of NAFLD progression. In this sense, in a recent study, we studied this possibility, but we failed to demonstrate any relationship between circulating SCFA levels and histological degrees of NAFLD in a cohort of patients with morbid obesity [6]. However, additional studies are necessary to accurately determine the specific role of SCFAs in NAFLD.

4.2 Bile acids

As previously mentioned, the gut-liver axis, which involves gut hormone release and the immune response, is essential to regulate systemic metabolism.

BAs participate in communication along this axis. They are steroid-derivative components of bile synthesized after cholesterol oxidation by enzymes present in hepatocytes, and they are involved in the absorption of lipids and vitamins in bile salt-dependent flow regulation. BAs participate in the digestion and solubilization of lipids and regulate hepatic glucose and inflammation [59, 60]. Moreover, they are capable of controlling their own synthesis through the activation of FXR [77, 78]. In addition, BAs act as signaling molecules that modulate several physiological processes, and GM dysbiosis can change BA pool characteristics through its effects on BA metabolism [78, 79].

The GM is a critical modulator of BA pool size and composition, and the process of dysbiosis could substantially alter concentrations of conjugated and/or secondary bile acids, as well as increase their synthesis.

Unmodified BAs, also called primary BAs (cholic acid (CA) and chenodeoxycholic acid (CDCA)), undergo a deconjugation process by GM components after reaching the colon and become secondary BAs, such as deoxycholic acid (DCA) and lithocholic acid (LCA); they can be transported again to the liver via the portal vein in a mechanism called "enterohepatic circulation." BAs prevent the overgrowth of bacteria in the gut to maintain gut homeostasis. This protective effect is mediated by their detergent properties and the activation of FXR, which protects the distal small intestine from bacterial proliferation. It is recognized that these circulating BAs, in addition to the abovementioned functions, can coordinate a wide number of pathways mediated by specific nuclear receptors (NRs) [60].

The increased intestinal permeability associated with BA modifications has been linked to metabolic endotoxemia, IR, and inflammatory cytokine release with enhanced proinflammatory signaling cascades, which are common findings in patients with NAFLD [59]. An increased level of BAs causes activation of the cell death pathway mediated by inflammatory and oxidative stress cascades in liver tissue [80, 81].

Regarding hepatic lipid metabolism, Watanabe et al. demonstrated that hepatic FXR activation mediated by BAs could induce the expression of the atypical NR small heterodimer partner (SHP), which promotes the inhibition of sterol-regulatory element-binding protein-1c (SREBP-1c), thus reducing hepatic synthesis of triglycerides. In addition, FXR can limit lipid accumulation in the liver by promoting fatty acid oxidation after the activation of peroxisome proliferator-activated receptor alpha (PPARα) and by the induction of plasma VLDL-triglyceride clearance [82–85]. FXR activation in the liver was also demonstrated to coordinate glucose homeostasis via the inhibition of gluconeogenesis and glycolysis. Interestingly, the activation of FXR in the intestine can generate crucial endocrine feedback regulation [86]. Experimental studies have demonstrated that intestinal dysbiosis can modulate the activity of FXR in the intestine, affecting lipid metabolism in the liver [4]. Specifically, FXR not only plays an important role in maintaining BA levels but also regulates glucose and lipid metabolism via different mechanisms, such as increasing insulin sensitivity, repressing hepatic gluconeogenic genes, and increasing hepatic glycogen synthesis [87, 88].

Previous investigations have demonstrated a BA level increase in the biological fluids of patients with NASH compared to that in the biological fluids of subjects with healthy livers and an evident association with intestinal dysbiosis [89–91]. Additionally, the levels of BAs have been correlated with histopathological features, such as the degree of hepatic steatosis, the presence of cellular ballooning, and the severity of fibrosis in patients with NASH [92]. These studies confirmed the disruption in BA homeostasis in NASH physiopathology [65] and the correlation of BAs with NASH severity parameters (portal inflammation, lobular inflammation, and hepatocyte ballooning) [93]. In children with NAFLD, changes in the circulating BA profile have also been reported. Troisi et al. demonstrated that serum BA levels decrease in early NAFLD and increase during progression to fibrosis in obese children. These authors postulated that BAs may have value as a noninvasive biomarker in pediatric NAFLD progression [83, 94]. In a previous study by our research group, we found that FXR jejunal expression was lower in NASH patients than in normal liver (NL) subjects; in regard to BAs, we also found that levels of glycolic acid (GCA), a primary BA, and DCA, a secondary BA, were significantly higher in NAFLD patients than in NL subjects [6].

Considering the numerous published experimental and clinical studies associating gut dysbiosis, BAs and NAFLD, it is expected that BAs could be proposed as potential noninvasive markers of the disease. For example, Svegliati-Baroni et al. specifically proposed DCA and LCA, which can only be produced by bacterial fermentation [95].

4.3 Bacterial components

The liver is exposed to potentially harmful substances derived from the gut, considered pathogen-associated molecular patterns (PAMPs), that include translocated bacteria, LPS, bacterial DNA, bacterial RNA, and endotoxins, which are potent inducers of tissue inflammation [41, 96]. These PAMPs might contribute to the pathogenesis of NAFLD by activating the innate immune system via TLRs, which recognize these gut-derived bacterial components. The healthy liver expresses low mRNA levels of TLRs (TLR1, TLR2, TLR4, TLR6, TLR7, TLR8, TLR9, and TLR10), implying a high tolerance of the liver to TLR ligands from the microbiota. The translocation of these bacterial components from the gut into the portal system is facilitated by intestinal barrier disruption due to GM dysbiosis [13, 96]. In this sense, there is evidence that dysbiosis causes permeability changes that increase portal levels of gut-derived TLR ligands (LPS or endotoxin), which further activate TLR4 on hepatic Kupffer and stellate cells [97]. LPS is the major structural component of gram-negative bacteria and the major component of endotoxin. LPS may be recognized by LPS-binding protein (LBP) in serum and is the major activator of the innate immune response [98]. Ruiz et al. indicated that the serum levels of LBP were increased in patients with obesity and NASH compared to those in patients with obesity and SS and the increased serum LBP level was correlated to an upregulated expression of TNF-α in liver tissue [99].

During TLR4 activation, the adaptor molecule myeloid differentiation factor 88 (MyD88) is activated, and the downstream signaling MyD88-dependent pathway results in the activation of necrosis factor kappa beta (NF-κB), leading to the expression of proinflammatory cytokines (TNF-α, IL-6, IL-8 and IL-12) and chemokines (interferon γ (IFN-γ) and monocyte chemotactic protein-1 (MCP-1)), promoting inflammation [68, 97]. There are several intracellular cascades involved in this process, generating oxidative stress, low-grade systemic inflammation, and hepatic injury [100]. In addition, TLR signaling can also lead to the production of inflammasomes in peripheral and parenchymal cells, which activate a variety of processes, including activation of caspase-1, resulting in cell death [101].

The inflammasome, which is a multimeric signaling platform that leads to the production of IL-18 and IL-1β through the NOD-like receptors pyrin domain-containing (NLRP3 and NLRP6), is activated by LPS derived from intestinal dysbiosis via TLR4 and TLR9 responses. Reports have associated inflammasome activation with the development of liver steatosis, inflammation, and fibrosis in NAFLD patients [102, 103].

It has been shown that TLR2, TLR4, and TLR9 play an important role in the development of NASH [104]. In addition, other studies have established that the increase in endotoxin levels is related to IL-1α and TNF-α production [105, 106]. In patients with NAFLD, gut permeability and SIBO due to intestinal dysbiosis have been associated with the severity of steatosis [107]. In biopsy-proven human NASH, plasma levels of IgG against endotoxin were found to be increased with NASH grade severity, suggesting the deleterious effect of chronic endotoxin exposure [108]. In our previous GM-derived metabolite study, we found overexpression of TLR9 jejunal expression in NAFLD subjects, which suggested the activation of the immune system during NAFLD progression [6]. Additionally, enhanced expression of TLR4, the release of IL-8, and high levels of LPS have been demonstrated in NAFLD patients [109, 110]. However, other reports did not reveal an association between endotoxemia and NAFLD progression, suggesting that endotoxemia may not be the only driver of disease development in all patients [111].

Multiple experimental studies have demonstrated that a high-fat diet can increase the proportion of LPS derived from the GM, and administration of endotoxin has been shown to induce IR and weight gain [99, 112]. On the other hand, some authors have recently proposed that the small intestine shields the liver from otherwise toxic fructose exposure via the GM [113].

There is a clear relation between gut dysbiosis, bacterial-derived components, the inflammatory response, and NAFLD; therefore, these bacterial mediators, especially circulating TLRs, might be used as potential noninvasive markers of disease progression.

4.4 Endogenous ethanol production

Intestinal dysbiosis increases endogenous ethanol production [111], which also affects gut permeability, disrupting intestinal tight junctions. This process allows endotoxins and ethanol to reach the liver and trigger the TLR response and inflammasome activation, contributing to liver damage [114]. In addition to the proinflammatory response, ethanol promotes oxidative stress and hepatocyte necrosis because of the formation of reactive oxygen and nitrogen species [94]. Endogenous ethanol inhibits the tricarboxylic acid cycle, thus increasing levels of acetate and thereby promoting triglyceride accumulation in hepatocytes [64]. Ethanol can also increase the activity of the enzyme cytochrome P450 2E1 (CYP2E1), which catalyzes the oxidation of ethanol but produces free radicals favoring oxidative damage, mitochondrial dysfunction, and liver inflammation [94, 115, 116].

Several studies have detected increased levels of non-dietary ethanol derived from bacteria in patients with obesity [111, 117] and in patients with NASH [111, 118, 119]. In this sense, Zhu et al. proposed that microbiomes rich in ethanol-producing *Escherichia* may be a risk factor for NAFLD progression [56]. *Escherichia*, *Bacteroides*, *Bifidobacterium*, and *Clostridium* can produce endogenous alcohol and generate significant ethanol-mediated liver damage [111]. Therefore, the production of endogenous ethanol by the GM may act as a hepatotoxin, contributing to the development of NAFLD and its progression to NASH [120]. In addition, children with NAFLD/NASH showed high levels of endogenous ethanol and LPS derived from the GM [111, 117, 121], confirming that endogenous ethanol might contribute to the pathogenesis of NAFLD and NASH.

Furthermore, Zhu et al. showed an increased abundance of alcohol-producing bacteria in NASH microbiomes, elevated blood-ethanol concentration in NASH patients, and the well-established role of alcohol metabolism in oxidative stress and liver inflammation [56]. In our previous GM-derived metabolite study, we found an interesting result about the higher circulating endogenous ethanol levels in NASH patients than in patients with SS. This fact suggested that circulating ethanol levels could distinguish between different degrees of liver damage. Moreover, in the same study, we evaluated the diagnostic efficacy of a biomarker panel including circulating ethanol, betaine,

GCA, and DCA levels as markers of NASH in a group of patients with liver histology indicative of NASH. A cutoff point and area under the curve were determined so that NASH could be diagnosed. The accuracy with which this panel discriminates NASH subjects from non-NASH subjects showed an area under the ROC curve (AUROC) of approximately 0.776 (0.632–0.921). Therefore, we concluded that the levels of certain circulating microbiota-related metabolites are associated with NAFLD severity and could be used as a "liquid biopsy" in the noninvasive diagnosis of NASH [6].

In summary, proinflammatory and prooxidative damage has been demonstrated as a result of endogenous ethanol in the liver, which might contribute to the pathogenesis of NAFLD, and previous reports may support its use as a noninvasive biomarker of disease progression.

4.5 Reduction of choline metabolism

Choline is an essential nutrient obtained through both dietary intake and endogenous synthesis and is an important constituent of the phospholipid membrane. The human GM actively metabolizes dietary components, including choline. Alterations in choline and phosphatidylcholine metabolism due to intestinal dysbiosis may have an impact on several physiological pathways, which could induce NAFLD. Choline deficiency prevents the synthesis and excretion of VLDL, leading to hepatic triglyceride accumulation and liver steatosis [122, 123]. In fact, the link between choline deficiency and the accumulation of hepatic lipids has been recognized for more than 50 years [124], leading to the establishment of choline-deficient diets to induce models of NAFLD in animals.

In addition, choline can be metabolized to its derivative trimethylamine (TMA) by the GM. TMA reaches the liver via portal circulation and is subsequently oxidized by hepatic flavin-containing monooxygenases in the liver, forming trimethylamine-N-oxide (TMAO), which is then released into blood circulation [125, 126]. Previous studies have revealed that TMAO may affect lipid absorption and cholesterol homeostasis and modulate glucose and lipid metabolism by decreasing the total BA pool size [122]. TMAO modulates glucose metabolism and increases IR in mice fed a high-fat diet [127]. TMAO also affects lipid absorption and cholesterol homeostasis by reducing the conversion of cholesterol into BAs [122].

A small number of human studies have shown that the consumption of a low-choline diet promotes fatty liver and liver damage [123, 128]. Other studies have pointed out that plasma-free choline levels are positively related to the severity of NAFLD, fibrosis, and NASH [129, 130].

On the other hand, in our previous research, we analyzed circulating levels of these choline metabolites according to hepatic histology and observed that levels of TMAO were significantly higher in NAFLD patients than in NL subjects [6], which correlates with the previous statement that serum TMAO levels are significantly higher in patients with NAFLD than in healthy people and correlates with the development and severity of NAFLD through different mechanisms: modulating glucose metabolism, promoting inflammation in adipose tissue, and influencing lipid absorption and cholesterol homeostasis [125, 129, 131].

In summary, the evidence has demonstrated that choline and TMAO are associated with the progression of NAFLD, indicating the potential use of these GM-derived mediators as markers of disease progression.

5. Gut microbiota as therapeutic target in NAFLD

Although there are no treatments to directly reverse steatosis, fibrosis, or liver damage, lifestyle changes and therapeutic strategies to treat other MS-related diseases, such as obesity, T2DM, or IR, could ameliorate NAFLD, avoiding its progression to NASH. Lifestyle intervention (diet and exercise), bariatric surgery, antidiabetic drugs, lipid-altering agents, and antihypertensive drugs can improve all of the features of NASH by ameliorating MS-related diseases [4]. Nevertheless, there is currently no specific treatment proven to be effective in treating NASH. Clarifying NAFLD risk factors could lead to more accurate prediction of disease progression and more effective treatments based on individualized drivers of disease [132]. The search for a possible therapy for NASH is focused on different pathways: metabolic targets, cell stress and apoptosis, immune targets, fibrosis, and GM modulation.

Currently, there are different mechanisms to manage NAFLD/NASH with metabolic targets focused on ameliorating other related diseases but also involved in NAFLD progression. Moreover, vitamin E acts as an antioxidant and hepatoprotective agent used to treat NASH (**Figure 2**).

On the other hand, there are many active studies and clinical trials focused on new therapeutic strategies with different pharmacological targets to avoid NAFLD and NASH progression. In this regard, PPAR agonists, antidiabetic drugs, FXR ligands, and anti-inflammatory and antiapoptotic agents can act as insulin sensitizers and improve the proinflammatory chronic state characteristic of NASH; antifibrotic agents can avoid NASH progression to fibrosis; and GM modulation can prevent the intestinal dysbiosis involved in NAFLD pathogenesis (**Figure 2**) [4, 7, 24, 133].

The key role that the GM plays in the progression of the disease opens the door to new ways of thinking about NASH prevention and treatment. The possibility of modulating the GM to treat NAFLD and NASH has gained interest in the potential use of probiotics, prebiotics, and antibiotics as effective treatments.

Probiotics are defined as viable microorganisms that when administered in adequate amounts, confer a health benefit to the host [134]. There are many mechanisms by which probiotics improve the GM and consequently ensure liver health (inhibition of intestinal bacterial enzymes, stimulation of host immunity, competition for limited nutrients, inhibition of bacterial mucosal adherence and epithelial

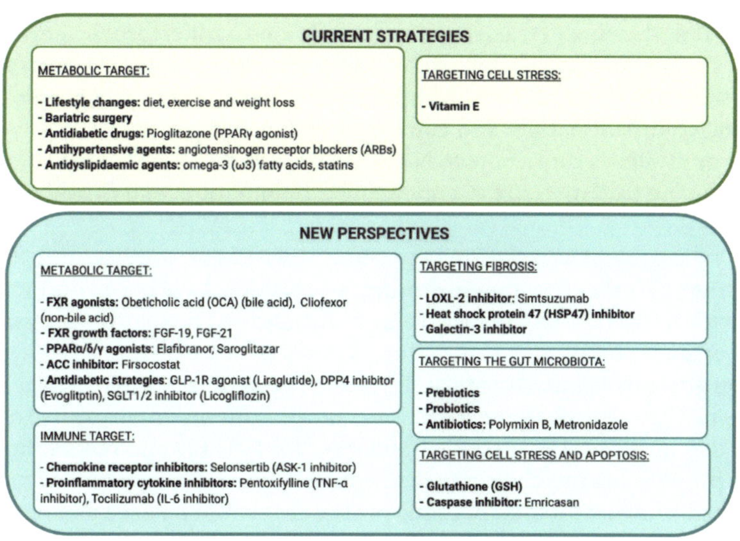

Figure 2.
Current and future treatment strategies to manage and treat NAFLD and NASH. Peroxisome proliferator-activated receptor (PPAR), farnesoid X receptor (FXR), farnesoid growth factor-19 (FGF-19), farnesoid growth factor-21 (FGF-21), acetyl-CoA carboxylase (ACC), glucagon-like peptide-1 receptor (GLP-1R), dipeptidyl peptidase-4 (DPP-4), sodium-glucose cotransporter 1/2 (SGLT-1/2), apoptosis signal-regulating kinase-1 (ASK-1), tumor necrosis factor alpha (TNF-α), interleukin-6 (IL-6), and lysyl oxidase-like 2 (LXL-2).

invasion, protection against intestinal permeability, and control of bacterial translocation from the gut to the portal vein circulation). The biological activity of probiotics depends on delivering anti-inflammatory mediators that downregulate proinflammatory cytokines [104]. Therefore, probiotic therapy offers an interesting approach to control hepatic injury and a low-grade proinflammatory state.

Another alternative is the use of prebiotic fiber, which is defined as an amount of nondigestible food ingredients that beneficially affect the host, by selectively stimulating the growth and/or activity of one or a limited number of bacteria in the colon [135]. The health effects of prebiotic fiber are related to improved glucoregulation and modified lipid metabolism as well as selective modulation of the GM. Some mechanisms have been proposed to explain the beneficial effects of prebiotics on the accumulation of triglycerides in the liver observed in animals, including reduced de novo fatty acid synthesis and SCFA production, body weight and fat loss, and improved glycemic control, GM modulation, and anti-inflammatory effects [13, 104]. These promising preliminary results strongly indicate the potential use of probiotics and prebiotics for the prevention or treatment of NASH.

Prophylactic use of antibiotics in patients with chronic liver diseases is an established method of preventing infections or innate immune dysfunction in acute liver failure (ALF) [13]. In addition, it has been demonstrated in animal and human models that the positive effect of polymyxin B and metronidazole in reducing the severity of NAFLD during total parenteral nutrition or after intestinal bypass could be interesting for their use to treat NAFLD [136, 137]. However, direct evidence is currently lacking, and thus, antibiotics cannot be routinely recommended to treat NASH, although further research is needed.

Overall, to date, there have been only a few studies concerning the use of probiotics, prebiotics, and antibiotics in humans; therefore, large-scale randomized controlled trials with histological endpoints are indicated.

6. Conclusions

Intestinal dysbiosis can trigger gut inflammation and increase the permeability of the intestinal epithelial barrier, exposing the gut-liver axis to GM-derived mediators of dysbiosis, such as bacterial components or metabolites, which may induce hepatotoxicity, inflammation, and consequently NAFLD progression. Gut-derived mediators of dysbiosis contribute to NAFLD progression by activating the immune system, inducing oxidative stress, enhancing inflammation, and finally promoting fibrogenesis.

Despite the evident association between GM dysbiosis, obesity, and NAFLD derived from several experimental studies, few studies have been conducted in patients with NAFLD to explore the role of GM-derived mediators of dysbiosis in the occurrence and progression of the disease. Additionally, few studies have focused on gut-derived mediators of dysbiosis as noninvasive markers of disease progression. The study of these mediators may provide an opportunity to develop a specific diagnostic and prognostic biomarker for NAFLD and NASH. In this sense, we propose the metabolomic study of these mediators and other metabolites involved to achieve a metabolomic profile that could be used as biomarkers for evaluating the status of NAFLD. On the other hand, some previous evidence has focused on GM modulation using probiotics, prebiotics, and antibiotics as therapeutic strategies to prevent or treat NAFLD and NASH, which is more uncertain and requires future research. In this sense, it remains important to promote study of GM targeting to find an effective treatment for NAFLD and overall for NASH.

Acknowledgements

This study was supported by the *Fondo de Investigación Sanitaria and Fondo Europeo de Desarrollo Regional* (FEDER, grant number PI16/00498, to Teresa Auguet), by funds from the *Agència de Gestió d'Ajuts Universitaris de Recerca* (AGAUR 2009 SGR 959 to Cristóbal Richart) and the *Grup de Recerca en Medicina Aplicada URV* (2016PFR-URV-B2-72 to Cristóbal Richart), and by the *Fundación Bionciencia*.

Conflict of interest

The authors declare no conflict of interest.

Abbreviations

NAFLD	non-alcoholic fatty liver disease
NASH	non-alcoholic steatohepatitis
GM	gut microbiota
T2DM	type 2 diabetes mellitus
IR	insulin resistance
MS	metabolic syndrome
LPS	lipopolysaccharides
SCFAs	short-chain fatty acids
BAs	bile acids
TMAO	trimethylamine-N-oxide
EtOH	ethanol
NAFL	non-alcoholic fatty liver
SS	simple steatosis
AST	aspartate aminotransferase
ALT	alanine aminotransferase
TLRs	toll-like receptors
SIBO	small intestinal bacterial overgrowth syndrome
FXR	farnesoid X receptor
VLDL	very-low density lipoprotein
GPR	G-protein coupled receptors
TNF-α	tumor necrosis factor alpha
IL	interleukin
GLP-1	glucagon-like peptide-1
AMPK	AMP-activated protein kinase
CA	cholic acid
CDCA	chenodeoxycholic acid
DCA	deoxycholic acid
LCA	lithocholic acid
NRs	nuclear receptors
SHP	small heterodimer partner
SREBP-1c	sterol-regulatory element-binding protein-1c
PPARα	proliferator-activated receptor alpha
NL	normal liver
GCA	glycolic acid
PAMPs	pathogen-associated molecular patterns
LPB	LPS-binding protein
NF-κB	necrosis factor-kappa beta

INF-γ	interferon gamma
MCP-1	monocyte chemotactic protein-1
NLRP	NOD-like receptors pyrin domain
CYP2E1	enzyme cytochrome P450 2E1
AUROC	area under the ROC curve
TMA	trimethylamine
FGF	farnesoid growth factor
ACC	acetyl-CoA carboxylase
DPP-4	dipeptidyl peptidase-4
SGLT-1/2	sodium-glucose cotransporter-1/2
ASK-1	apoptosis signal-regulating kinase-1
LXL-2	lysyl oxidase-like-2
ALF	acute liver failure

Author details

Teresa Auguet[1*], Laia Bertran[2] and Jessica Binetti[1]

1 Internal Medicine Department, University Hospital of Tarragona Joan XXIII, Rovira and Virgili University, Institut d'Investigació Sanitària Pere Virgili, Tarragona, Spain

2 Medicine and Surgery Department, Rovira and Virgili University, Institut d'Investigació Sanitària Pere Virgili, Tarragona, Spain

*Address all correspondence to: tauguet.hj23.ics@gencat.cat

References

[1] Cotter TG, Rinella M. NAFLD 2020: The state of the disease. Gastroenterology. 2020;**158**:1851-1864. DOI: 10.1053/j.gastro.2020.01.052

[2] Younossi ZM et al. The economic and clinical burden of nonalcoholic fatty liver disease in the United States and Europe. Hepatology. 2016;**64**(5): 1577-1586. DOI: 10.1002/hep.28785

[3] Ekstedt M et al. Long-term follow-up of patients with NAFLD and elevated liver enzymes. Hepatology. 2006;**44**(4):865-873. DOI: 10.1002/hep.21327

[4] Chalasani N et al. The diagnosis and management of non-alcoholic fatty liver disease: Practice guideline by the American Association for the Study of Liver Diseases, American College of Gastroenterology and the American Gastroenterological Association. The American Journal of Gastroenterology. 2012;**107**(6):811-826. DOI: 10.1038/ajg.2012.128

[5] El-Badry AM et al. Assessment of hepatic steatosis by expert pathologists: The end of a gold standard. Annals of Surgery. 2009;**250**(5):691-696. DOI: 10.1097/SLA.0b013e3181bcd6dd

[6] Aragonès G et al. Circulating microbiota-derived metabolites: A "liquid biopsy?". International Journal of Obesity. 2019;**44**:875-885. DOI: 10.1038/s41366-019-0430-0

[7] Carter-Kent C, Zein NN, Feldstein AE. Cytokines in the pathogenesis of fatty liver and disease progression to steatohepatitis: Implications for treatment. The American Journal of Gastroenterology. 2008;**103**(4):1036-1042. DOI: 10.1111/j.1572-0241.2007.01709.x

[8] Aron-Wisnewsky J, Gaborit B, Dutour A, Clement K. Gut microbiota and non-alcoholic fatty liver disease: New insights. Clinical Microbiology and Infection. 2013;**19**(4):338-348. DOI: 10.1111/1469-0691.12140

[9] Gangarapu V, Yildiz K, Tuzun Ince A, Baysal B. Role of gut microbiota: Obesity and NAFLD. The Turkish Journal of Gastroenterology. 2014;**25**(2):133-140. DOI: 10.5152/tjg.2014.7886

[10] Shen F, Zheng R-D, Sun X-Q, Ding W-J, Wang X-Y, Fan J-G. Gut microbiota dysbiosis in patients with non-alcoholic fatty liver disease. Hepatobiliary & Pancreatic Diseases International. 2017;**16**(4):375-381. DOI: 10.1016/S1499-3872(17)60019-5

[11] Janssen S, Depoortere I. Nutrient sensing in the gut: New roads to therapeutics? Trends in Endocrinology and Metabolism. 2013;**24**(2):92-100. DOI: 10.1016/j.tem.2012.11.006

[12] Okubo H et al. Roles of gut-derived secretory factors in the pathogenesis of non-alcoholic fatty liver disease and their possible clinical applications. International Journal of Molecular Sciences. 2018;**19**(10):3064. DOI: 10.3390/ijms19103064

[13] Imajo K, Yoneda M, Ogawa Y, Wada K, Nakajima A. Microbiota and nonalcoholic steatohepatitis. Seminars in Immunopathology. 2014;**36**(1):115-132. DOI: 10.1007/s00281-013-0404-6

[14] Machado M, Marques-Vidal P, Cortez-Pinto H. Hepatic histology in obese patients undergoing bariatric surgery. Journal of Hepatology. 2006;**45**(4):600-606. DOI: 10.1016/j.jhep.2006.06.013

[15] Lewis JR, Mohanty SR. Nonalcoholic fatty liver disease: A review and update. Digestive Diseases and Sciences. 2010;**55**(3):560-578. DOI: 10.1007/s10620-009-1081-0

[16] Makoto N et al. Evaluation of fatty acid metabolism-related gene expression in nonalcoholic fatty liver disease. International Journal of Molecular Medicine. 2005;**16**(4):631-635

[17] Younossi Z, Henry L. Contribution of alcoholic and nonalcoholic fatty liver disease to the burden of liver-related morbidity and mortality. Gastroenterology. 2016;**150**(8):1778-1785. DOI: 10.1053/j.gastro.2016.03.005

[18] Abd El-Kader SM, El-Den Ashmawy EMS. Non-alcoholic fatty liver disease: The diagnosis and management. World Journal of Hepatology. 2015;7(6):846-858. DOI: 10.4254/wjh.v7.i6.846

[19] Wainwright P, Byrne CD. Bidirectional relationships and disconnects between NAFLD and features of the metabolic syndrome. International Journal of Molecular Sciences. 2016;**17**(3). DOI: 10.3390/ijms17030367

[20] Marchesini G et al. Nonalcoholic fatty liver, steatohepatitis, and the metabolic syndrome. Hepatology. 2003;**37**(4):917-923. DOI: 10.1053/jhep.2003.50161

[21] Adams LA, Waters OR, Knuiman MW, Elliott RR, Olynyk JK. NAFLD as a risk factor for the development of diabetes and the metabolic syndrome: An eleven-year follow-up study. The American Journal of Gastroenterology. 2009;**104**(4):861-867. DOI: 10.1038/ajg.2009.67

[22] Yan E, Durazo F, Tong M, Hong K. Nonalcoholic fatty liver disease: Pathogenesis, identification, progression, and management. Nutrition in Clinical Care. 2007;**65**(8):376-384. DOI: 10.1301/nr.2007.aug.376

[23] Mofrad P et al. Clinical and histologic spectrum of nonalcoholic fatty liver disease associated with normal ALT values. Hepatology. 2003;**37**(6):1286-1292. DOI: 10.1053/jhep.2003.50229

[24] Federico A, Trappoliere M, Loguercio C. Treatment of patients with non-alcoholic fatty liver disease: Current views and perspectives. Digestive and Liver Disease. 2006;**38**(11):789-801. DOI: 10.1016/j.dld.2006.04.009

[25] Pacifico L, Celestre M, Anania C, Paolantonio P, Chiesa C, Laghi A. MRI and ultrasound for hepatic fat quantification: Relationships to clinical and metabolic characteristics of pediatric nonalcoholic fatty liver disease. Acta Paediatrica, International Journal of Paediatrics. 2007;**96**(4):542-547. DOI: 10.1111/j.1651-2227.2007.00186.x

[26] Chalasani N et al. The diagnosis and management of nonalcoholic fatty liver disease: Practice guidance from the American Association for the Study of Liver Diseases. Hepatology. 2018;**67**(1):328-357. DOI: 10.1002/hep.29367

[27] Kleiner DE et al. Design and validation of a histological scoring system for nonalcoholic fatty liver disease. Hepatology. 2005;**41**(6):1313-1321. DOI: 10.1002/hep.20701

[28] Buzzetti E, Pinzani M, Tsochatzis EA. The multiple-hit pathogenesis of non-alcoholic fatty liver disease (NAFLD). Metabolism. 2016;**65**(8):1038-1048. DOI: 10.1016/j.metabol.2015.12.012

[29] Shen C et al. The relationship between hepatic resistin overexpression and inflammation in patients with nonalcoholic steatohepatitis. BMC Gastroenterology. 2014;**14**(1):1-8. DOI: 10.1186/1471-230X-14-39

[30] Tilg H, Moschen AR. Evolution of inflammation in nonalcoholic

fatty liver disease: The multiple parallel hits hypothesis. Hepatology. 2010;**52**(5):1836-1846. DOI: 10.1002/hep.24001

[31] Liaskou E, Wilson DV, Oo YH. Innate immune cells in liver inflammation. Mediators of Inflammation. 2012;**949157**. DOI: 10.1155/2012/949157

[32] Wan X, Xu C, Yu C, Li Y. Role of NLRP3 inflammasome in the progression of NAFLD to NASH. Canadian Journal of Gastroenterology & Hepatology. 2016;**2016**(4):6489012. DOI: 10.1155/2016/6489012

[33] Abdel-Misih SRZ, Bloomston M. Liver anatomy. The Surgical Clinics of North America. 2010;**90**(4):643-653. DOI: 10.1016/j.suc.2010.04.017

[34] Wang B, Yao M, Lv L, Ling Z, Li L. The human microbiota in health and disease. Engineering. 2017;**3**(1): 71-82. DOI: 10.1016/J.ENG.2017.01.008

[35] Gill SRG et al. Metagenomic analysis of the human distal gut microbiome. Science. 2006;**312**(5778):1355-1359. DOI: 10.1126/science.1124234. Metagenomic

[36] Seksik P, Landman C. Understanding microbiome data: A primer for clinicians. Digestive Diseases. 2015;**33**(suppl 1):11-16. DOI: 10.1159/000437034

[37] Xu J, Gordon JI. Honor thy symbionts. Proceedings of the National Academy of Sciences of the United States of America. 2003;**100**(18):10452-10459. DOI: 10.1073/pnas.1734063100

[38] Han R, Ma J, Li H. Mechanistic and therapeutic advances in non-alcoholic fatty liver disease by targeting the gut microbiota. Frontiers in Medicine. 2018;**12**(6):645-657. DOI: 10.1007/s11684-018-0645-9

[39] Lozupone CA, Stombaugh JI, Gordon JI, Jansson JK, Knight R. Diversity, stability and resilience of the human gut microbiota. Nature. 2012;**489**(7415):220-230. DOI: 10.1038/nature11550

[40] Faith JJ et al. The long-term stability of the human gut microbiota. Science. 2013;**341**(6141):1237439-1237439. DOI: 10.1126/science.1237439

[41] Leung C, Rivera L, Furness JB, Angus PW. The role of the gut microbiota in NAFLD. Nature Reviews. Gastroenterology & Hepatology. 2016;**13**(7):412-425. DOI: 10.1038/nrgastro.2016.85

[42] Chassaing B, Gewirtz AT. Gut microbiota, low-grade inflammation, and metabolic syndrome. Toxicologic Pathology. 2014;**42**(1):49-53. DOI: 10.1177/0192623313508481

[43] Acharya C, Bajaj JS. Gut microbiota and complications of liver disease. Gastroenterology Clinics of North America. 2017;**46**(1):155-169. DOI: 10.1016/j.gtc.2016.09.013

[44] Bäckhed F et al. The gut microbiota as an environmental factor that regulates fat storage. Proceedings of the National Academy of Sciences of the United States of America. 2004;**101**(44):15718-15723. DOI: 10.1073/pnas.0407076101

[45] Brandl K, Kumar V, Eckmann L. Gut-liver axis at the frontier of host-microbial interactions. American Journal of Physiology. Gastrointestinal and Liver Physiology. 2017;**312**(5):G413-G419. DOI: 10.1152/ajpgi.00361.2016

[46] Szabo G. Gut-liver axis in alcoholic liver disease. Gastroenterology. 2015;**148**(1):30-36. DOI: 10.1053/j.gastro.2014.10.042

[47] Manfredo Vieira S et al. Translocation of a gut pathobiont drives

autoimmunity in mice and humans. Science. 2018;**359**(6380):1156-1161. DOI: 10.1126/science.aar7201

[48] Thaiss CA et al. Hyperglycemia drives intestinal barrier dysfunction and risk for enteric infection. Science. 2018;**359**(6382):1376-1383. DOI: 10.1126/science.aar3318

[49] Chu H, Duan Y, Yang L, Schnabl B. Small metabolites, possible big changes: A microbiota-centered view of non-alcoholic fatty liver disease. Gut. 2019;**68**(2):359-370. DOI: 10.1136/gutjnl-2018-316307

[50] Schnabl B, Brenner DA. Interactions between the intestinal microbiome and liver diseases. Gastroenterology. 2014;**146**(6):1513-1524. DOI: 10.1053/j.gastro.2014.01.020

[51] Langille MGI et al. Predictive functional profiling of microbial communities using 16S rRNA marker gene sequences. Nature Biotechnology. 2013;**31**(9):814-821. DOI: 10.1038/nbt.2676

[52] Boursier J et al. The severity of nonalcoholic fatty liver disease is associated with gut dysbiosis and shift in the metabolic function of the gut microbiota. Hepatology. 2016;**63**(3):764-775. DOI: 10.1002/hep.28356

[53] Mouzaki M et al. Intestinal microbiota in patients with nonalcoholic fatty liver disease. Hepatology. 2013;**58**(1):120-127. DOI: 10.1002/hep.26319

[54] Fukui H. Role of gut dysbiosis in liver diseases: What have we learned so far? Diseases. 2019;**7**(4):58. DOI: 10.3390/diseases7040058

[55] Jiang W et al. Dysbiosis gut microbiota associated with inflammation and impaired mucosal immune function in intestine of humans with non-alcoholic fatty liver disease. Scientific Reports. 2015;**5**:1-7. DOI: 10.1038/srep08096

[56] Zhu L et al. Characterization of gut microbiomes in nonalcoholic steatohepatitis (NASH) patients: A connection between endogenous alcohol and NASH. Hepatology. 2013;**57**(2):601-609. DOI: 10.1002/hep.26093

[57] Sobhonslidsuk A et al. The association of gut microbiota with nonalcoholic steatohepatitis in thais. BioMed Research International. 2018;**2018**:9340316. DOI: 10.1155/2018/9340316

[58] Qin N et al. Alterations of the human gut microbiome in liver cirrhosis. Nature. 2014;**513**(7516):59-64. DOI: 10.1038/nature13568

[59] Aragonès G, González-garc S, Aguilar C, Richart C, Auguet T. Gut microbiota-derived mediators as potential markers in nonalcoholic fatty liver disease. BioMed Research International. 2019;**2019**:8507583. DOI: 10.1155/2019/8507583

[60] Marra F, Svegliati-Baroni G. Lipotoxicity and the gut-liver axis in NASH pathogenesis. Journal of Hepatology. 2018;**68**(2):280-295. DOI: 10.1016/j.jhep.2017.11.014

[61] Chakraborti CK. New-found link between microbiota and obesity. World Journal of Gastrointestinal Pathophysiology. 2015;**6**(4):110-119. DOI: 10.4291/wjgp.v6.i4.110

[62] Arslan N. Obesity, fatty liver disease and intestinal microbiota. World Journal of Gastroenterology. 2014;**20**(44):16452. DOI: 10.3748/wjg.v20.i44.16452

[63] Loomba R et al. Gut microbiome-based metagenomic signature for non-invasive detection of advanced fibrosis in human nonalcoholic

fatty liver disease. Cell Metabolism. 2017;**25**(5):1054.e5-1062.e5. DOI: 10.1016/j.cmet.2017.04.001

[64] Michail S et al. Altered gut microbial energy and metabolism in children with non-alcoholic fatty liver disease. FEMS Microbiology Ecology. 2015;**91**(2):1-9. DOI: 10.1093/femsec/fiu002

[65] Ferslew BC et al. Altered bile acid metabolome in patients with nonalcoholic steatohepatitis. Digestive Diseases and Sciences. 2015;**60**(11):3318-3328. DOI: 10.1007/s10620-015-3776-8

[66] Aranha MM et al. Bile acid levels are increased in the liver of patients with steatohepatitis. European Journal of Gastroenterology & Hepatology. 2008;**20**(6):519-525. DOI: 10.1097/MEG.0b013e3282f4710a

[67] Kalhan SC et al. Plasma metabolomic profile in nonalcoholic fatty liver disease. Metabolism. 2011;**60**(3):404-413. DOI: 10.1016/j.metabol.2010.03.006

[68] Mouzaki M et al. Bile acids and dysbiosis in non-alcoholic fatty liver disease. PLoS One. 2016;**11**(5):e0151829. DOI: 10.1371/journal.pone.0151829

[69] Lu LP et al. Serum bile acid level and fatty acid composition in Chinese children with non-alcoholic fatty liver disease. Journal of Digestive Diseases. 2017;**18**(8):461-471. DOI: 10.1111/1751-2980.12494

[70] Shanab AA et al. Small intestinal bacterial overgrowth in nonalcoholic steatohepatitis: Association with toll-like receptor 4 expression and plasma levels of interleukin 8. Digestive Diseases and Sciences. 2011;**56**(5):1524-1534. DOI: 10.1007/s10620-010-1447-3

[71] Engstler AJ et al. Insulin resistance alters hepatic ethanol metabolism: Studies in mice and children with non-alcoholic fatty liver disease. Gut. 2016;**65**(9):1564-1571. DOI: 10.1136/gutjnl-2014-308379

[72] Imajo K et al. Plasma free choline is a novel non-invasive biomarker for early-stage non-alcoholic steatohepatitis: A multi-center validation study. Hepatology Research. 2012;**42**(8):757-766. DOI: 10.1111/j.1872-034X.2012.00976.x

[73] Park J et al. Short-chain fatty acids induce both effector and regulatory T cells by suppression of histone deacetylases and regulation of the mTOR-S6K pathway. Mucosal Immunology. 2015;**8**(1):80-93. DOI: 10.1038/mi.2014.44

[74] Pierri L et al. Gut microbiota (GM) and gut-liver axis (GLA) components in obesity related NAFLD: A pediatric pilot study. Digestive and Liver Disease. 2015;**47**(Suppl. 4):e253. DOI: 10.1016/j.dld.2015.07.091

[75] Guerrerio AL et al. Choline intake in a large cohort of patients with nonalcoholic fatty liver disease. The American Journal of Clinical Nutrition. 2012;**95**(4):892-900. DOI: 10.3945/ajcn.111.020156

[76] Chen Y et al. Associations of gut-flora-dependent metabolite trimethylamine-N-oxide, betaine and choline with non-alcoholic fatty liver disease in adults. Scientific Reports. 2016;**6**(1):19076. DOI: 10.1038/srep19076

[77] Alves-Bezerra M, Cohen DE. Triglyceride metabolism in the liver. Comprehensive Physiology. 2017;**8**(1):1-8. DOI: 10.1002/cphy.c170012

[78] den Besten G et al. Gut-derived short-chain fatty acids are vividly assimilated into host carbohydrates and lipids. American Journal of Physiology.

Gastrointestinal and Liver Physiology. 2013;**305**(12):G900-G910. DOI: 10.1152/ajpgi.00265.2013

[79] Vinolo MAR, Rodrigues HG, Hatanaka E, Hebeda CB, Farsky SHP, Curi R. Short-chain fatty acids stimulate the migration of neutrophils to inflammatory sites. Clinical Science (London, England). 2009;**117**(9): 331-338. DOI: 10.1042/CS20080642

[80] Tan J, McKenzie C, Potamitis M, Thorburn AN, Mackay CR, Macia L. The role of short-chain fatty acids in health and disease. Advances in Immunology. 2014;**121**:91-119

[81] Koh A, De Vadder F, Kovatcheva-Datchary P, Bäckhed F. From dietary fiber to host physiology: Short-chain fatty acids as key bacterial metabolites. Cell. 2016;**165**(6):1332-1345. DOI: 10.1016/j.cell.2016.05.041

[82] Watanabe M et al. Bile acids lower triglyceride levels via a pathway involving FXR, SHP, and SREBP-1c. The Journal of Clinical Investigation. 2004;**113**(10):1408-1418. DOI: 10.1172/JCI21025

[83] Pineda Torra I, Claudel T, Duval C, Kosykh V, Fruchart J-C, Staels B. Bile acids induce the expression of the human peroxisome proliferator-activated receptor alpha gene via activation of the farnesoid X receptor. Molecular Endocrinology. 2003;**17**(2):259-272. DOI: 10.1210/me.2002-0120

[84] Savkur RS, Bramlett KS, Michael LF, Burris TP. Regulation of pyruvate dehydrogenase kinase expression by the farnesoid X receptor. Biochemical and Biophysical Research Communications. 2005;**329**(1):391-396. DOI: 10.1016/j.bbrc.2005.01.141

[85] Fuchs CD, Traussnigg SA, Trauner M. Nuclear receptor modulation for the treatment of nonalcoholic fatty liver disease. Seminars in Liver Disease. 2016;**36**(1):69-86. DOI: 10.1055/s-0036-1571296

[86] Inagaki T et al. Fibroblast growth factor 15 functions as an enterohepatic signal to regulate bile acid homeostasis. Cell Metabolism. 2005;**2**(4):217-225. DOI: 10.1016/j.cmet.2005.09.001

[87] Hylemon PB, Zhou H, Pandak WM, Ren S, Gil G, Dent P. Bile acids as regulatory molecules. Journal of Lipid Research. 2009;**50**(8):1509-1520. DOI: 10.1194/jlr.R900007-JLR200

[88] Allen K, Jaeschke H, Copple BL. Bile acids induce inflammatory genes in hepatocytes: A novel mechanism of inflammation during obstructive cholestasis. The American Journal of Pathology. 2011;**178**(1):175-186. DOI: 10.1016/j.ajpath.2010.11.026

[89] Svegliati-Baroni G et al. Glucagon-like peptide-1 receptor activation stimulates hepatic lipid oxidation and restores hepatic signalling alteration induced by a high-fat diet in nonalcoholic steatohepatitis. Liver International. 2011;**31**(9):1285-1297. DOI: 10.1111/j.1478-3231.2011.02462.x

[90] Gao Z et al. Butyrate improves insulin sensitivity and increases energy expenditure in mice. Diabetes. 2009;**58**(7):1509-1517. DOI: 10.2337/db08-1637

[91] Peng L, Li Z-R, Green RS, Holzman IR, Lin J. Butyrate enhances the intestinal barrier by facilitating tight junction assembly via activation of AMP-activated protein kinase in Caco-2 cell monolayers. The Journal of Nutrition. 2009;**139**(9):1619-1625. DOI: 10.3945/jn.109.104638

[92] Tao R et al. Deacetylase inhibition promotes the generation and function of regulatory T cells. Nature Medicine. 2007;**13**(11):1299-1307. DOI: 10.1038/nm1652

[93] Fuchs C, Claudel T, Trauner M. Bile acid-mediated control of liver triglycerides. Seminars in Liver Disease. 2013;**33**(4):330-342. DOI: 10.1055/s-0033-1358520

[94] Troisi J et al. Urinary metabolomics in pediatric obesity and NAFLD identifies metabolic pathways/ metabolites related to dietary habits and gut-liver axis perturbations. Nutrients. 2017;**9**:485. DOI: 10.3390/nu9050485

[95] Svegliati-Baroni G et al. Bile acids induce hepatic stellate cell proliferation via activation of the epidermal growth factor receptor. Gastroenterology. 2005;**128**(4):1042-1055

[96] Miele L et al. Increased intestinal permeability and tight junction alterations in nonalcoholic fatty liver disease. Hepatology. 2009;**49**(6): 1877-1887. DOI: 10.1002/hep.22848

[97] Giorgio V et al. Intestinal permeability is increased in children with non-alcoholic fatty liver disease, and correlates with liver disease severity. Digestive and Liver Disease. 2014;**46**(6):556-560. DOI: 10.1016/j. dld.2014.02.010

[98] Liu Q et al. Role and effective therapeutic target of gut microbiota in NAFLD/NASH. Experimental and Therapeutic Medicine. 2019;**18**: 1935-1944. DOI: 10.3892/etm.2019.7781

[99] Ruiz AG et al. Lipopolysaccharide-binding protein plasma levels and liver TNF-alpha gene expression in obese patients: Evidence for the potential role of endotoxin in the pathogenesis of non-alcoholic steatohepatitis. Obesity Surgery. 2007;**17**(10):1374-1380. DOI: 10.1007/s11695-007-9243-7

[100] Puri P et al. The circulating microbiome signature and inferred functional metagenomics in alcoholic hepatitis. Hepatology. 2018;**67**(4): 1284-1302. DOI: 10.1002/hep.29623

[101] Poeta M, Pierri L, Vajro P. Gut-liver axis derangement in non-alcoholic fatty liver disease. Children (Basel, Switzerland). 2017;**4**(8). DOI: 10.3390/ children4080066

[102] De Fabiani E, Mitro N, Gilardi F, Caruso D, Galli G, Crestani M. Coordinated control of cholesterol catabolism to bile acids and of gluconeogenesis via a novel mechanism of transcription regulation linked to the fasted-to-fed cycle. The Journal of Biological Chemistry. 2003;**278**(40):39124-39132. DOI: 10.1074/jbc.M305079200

[103] Zhang Y et al. Activation of the nuclear receptor FXR improves hyperglycemia and hyperlipidemia in diabetic mice. Proceedings of the National Academy of Sciences of the United States of America. 2006;**103**(4):1006-1011. DOI: 10.1073/ pnas.0506982103

[104] Compare D et al. Gut-liver axis: The impact of gut microbiota on non alcoholic fatty liver disease. Nutrition, Metabolism, and Cardiovascular Diseases. 2012;**22**(6):471-476. DOI: 10.1016/j.numecd.2012.02.007

[105] Mencin A, Kluwe J, Schwabe RF. Toll-like receptors as targets in chronic liver diseases. Gut. 2009;**58**(5):704-720. DOI: 10.1136/gut.2008.156307

[106] Singh R et al. Status of bacterial colonization, toll-like receptor expression and nuclear factor-kappa B activation in normal and diseased human livers. Clinical Immunology. 2011;**138**(1):41-49. DOI: 10.1016/j. clim.2010.09.006

[107] Samuel BS et al. Effects of the gut microbiota on host adiposity are modulated by the short-chain fatty-acid binding G protein-coupled receptor, Gpr41. Proceedings of the National Academy of Sciences of the United States of America.

2008;**105**(43):16767-16772. DOI: 10.1073/pnas.0808567105

[108] Dixon LJ, Flask CA, Papouchado BG, Feldstein AE, Nagy LE. Caspase-1 as a central regulator of high fat diet-induced non-alcoholic steatohepatitis. PLoS One. 2013;**8**(2):e56100. DOI: 10.1371/journal.pone.0056100

[109] Musso G, Gambino R, Cassader M. Obesity, diabetes, and gut microbiota: The hygiene hypothesis expanded? Diabetes Care. 2010;**33**(10):2277-2284. DOI: 10.2337/dc10-0556

[110] Henao-Mejia J et al. Inflammasome-mediated dysbiosis regulates progression of NAFLD and obesity. Nature. 2012;**482**(7384):179-185

[111] Pang J et al. Significant positive association of endotoxemia with histological severity in 237 patients with non-alcoholic fatty liver disease. Alimentary Pharmacology & Therapeutics. 2017;**46**(2):175-182. DOI: 10.1111/apt.14119

[112] Poniachik J et al. Increased production of IL-1alpha and TNF-alpha in lipopolysaccharide-stimulated blood from obese patients with non-alcoholic fatty liver disease. Cytokine. 2006;**33**(5):252-257. DOI: 10.1016/j.cyto.2006.02.006

[113] Verdam FJ, Rensen SS, Driessen A, Greve JW, Buurman WA. Novel evidence for chronic exposure to endotoxin in human nonalcoholic steatohepatitis. Journal of Clinical Gastroenterology. 2011;**45**(2):149-152. DOI: 10.1097/MCG.0b013e3181e12c24

[114] Guercio NS, Di SM, Pierri L, Troisi J, Poeta M. Multiple gut-liver axis abnormalities in children with obesity with and without hepatic involvement. Pediatric Obesity. 2017;**12**:446-452. DOI: 10.1111/ijpo.12164

[115] Mehta NN et al. Experimental endotoxemia induces adipose inflammation and insulin resistance in humans. Diabetes. 2010;**59**(1):172-181. DOI: 10.2337/db09-0367

[116] Cani PD et al. Metabolic endotoxemia initiates obesity and insulin resistance. Diabetes. 2007;**56**(7):1761-1772. DOI: 10.2337/db06-1491

[117] Jang C et al. The small intestine converts dietary fructose into glucose and organic acids. Cell Metabolism. 2018;**27**(2):351.e3-361.e3. DOI: 10.1016/j.cmet.2017.12.016

[118] Parlesak A, Schäfer C, Schütz T, Bode JC, Bode C. Increased intestinal permeability to macromolecules and endotoxemia in patients with chronic alcohol abuse in different stages of alcohol-induced liver disease. Journal of Hepatology. 2000;**32**(5):742-747. DOI: 10.1016/s0168-8278(00)80242-1

[119] Lieber CS. Microsomal ethanol-oxidizing system (MEOS): The first 30 years (1968-1998)—A review. Alcoholism, Clinical and Experimental Research. 1999;**23**(6):991-1007. DOI: 10.1111/j.1530-0277.1999.tb04217.x

[120] Raucy JL, Lasker J, Ozaki K, Zoleta V. Regulation of CYP2E1 by ethanol and palmitic acid and CYP4A11 by clofibrate in primary cultures of human hepatocytes. Toxicological Sciences. 2004;**79**(2):233-241. DOI: 10.1093/toxsci/kfh126

[121] Elamin E, Masclee A, Troost F, Dekker J, Jonkers D. Cytotoxicity and metabolic stress induced by acetaldehyde in human intestinal LS174T goblet-like cells. American Journal of Physiology. Gastrointestinal and Liver Physiology. 2014;**307**(3):G286-G294. DOI: 10.1152/ajpgi.00103.2014

[122] Nair S, Cope K, Risby TH, Diehl AM, Terence RH. Obesity and female gender increase breath ethanol concentration: Potential implications for the pathogenesis of nonalcoholic steatohepatitis. The American Journal of Gastroenterology. 2001;**96**(4):1200-1204. DOI: 10.1111/j.1572-0241.2001.03702.x

[123] Baker SS, Baker RD, Liu W, Nowak NJ, Zhu L. Role of alcohol metabolism in non-alcoholic steatohepatitis. PLoS One. 2010;**5**(3):e9570. DOI: 10.1371/journal.pone.0009570

[124] Bashiardes S, Shapiro H, Rozin S, Shibolet O, Elinav E. Non-alcoholic fatty liver and the gut microbiota. Molecular Metabolism. 2016;**5**(9):782-794. DOI: 10.1016/j.molmet.2016.06.003

[125] Koeth RA et al. Intestinal microbiota metabolism of L-carnitine, a nutrient in red meat, promotes atherosclerosis. Nature Medicine. 2013;**19**(5):576-585. DOI: 10.1038/nm.3145

[126] Nakamura T, Nakamura S, Karoji N, Aikawa T, Suzuki O. Hepatic function tests in heavy drinkers among workmen. The Tohoku Journal of Experimental Medicine. 1967;**93**(3):219-226. DOI: 10.1620/tjem.93.219

[127] Wang B et al. Altered fecal microbiota correlates with liver biochemistry in nonobese patients with non-alcoholic fatty liver disease. Scientific Reports. 2016;**6**(1):32002. DOI: 10.1038/srep32002

[128] Bennett BJ et al. Trimethylamine-N-oxide, a metabolite associated with atherosclerosis, exhibits complex genetic and dietary regulation. Cell Metabolism. 2013;**17**(1):49-60. DOI: 10.1016/j.cmet.2012.12.011

[129] Gao X, Liu X, Xu J, Xue C, Xue Y, Wang Y. Dietary trimethylamine N-oxide exacerbates impaired glucose tolerance in mice fed a high fat diet. Journal of Bioscience and Bioengineering. 2014;**118**(4):476-481. DOI: 10.1016/j.jbiosc.2014.03.001

[130] Corbin KD, Zeisel SH. Choline metabolism provides novel insights into nonalcoholic fatty liver disease and its progression. Current Opinion in Gastroenterology. 2012;**28**(2):159-165. DOI: 10.1097/MOG.0b013e32834e7b4b

[131] Chen YM et al. Associations of gut-flora-dependent metabolite trimethylamine-N-oxide, betaine and choline with non-alcoholic fatty liver disease in adults. Scientific Reports. 2016;**6**:1-9. DOI: 10.1038/srep19076

[132] Friedman SL, Neuschwander-Tetri BA, Rinella M, Sanyal AJ. Mechanisms of NAFLD development and therapeutic strategies. Nature Medicine. 2018;**24**(7):908-922. DOI: 10.1038/s41591-018-0104-9

[133] Sumida Y, Yoneda M. Current and future pharmacological therapies for NAFLD/NASH. Journal of Gastroenterology. 2018;**53**(3):362-376. DOI: 10.1007/s00535-017-1415-1

[134] Minocha A. Probiotics for preventive health. Nutrition in Clinical Practice. 2009;**24**:227-241. DOI: 10.1177/0884533608331177

[135] Parnell JA, Raman M, Rioux KP, Reimer RA. The potential role of prebiotic fibre for treatment and management of non-alcoholic fatty liver disease and associated obesity and insulin resistance. Liver International. 2012;**32**(5):701-711. DOI: 10.1111/j.1478-3231.2011.02730.x

[136] Pappo I, Becovier H, Berry EM, Freund HR. Polymyxin B reduces cecal flora, TNF production and hepatic

steatosis during total parenteral nutrition in the rat. The Journal of Surgical Research. 1991;**51**(2):106-112. DOI: 10.1016/0022-4804(91)90078-Z

[137] Drenick EJ, Fisler J, Johnson D. Hepatic steatosis after intestinal bypass—Prevention and reversal by metronidazole, irrespective of protein-calorie malnutrition. Gastroenterology. 1982;**82**(3):535-548. DOI: 10.1016/S0016-5085(82)80403-4

Chapter 4

Skin and Gut Microbiota in Psoriasis: A Systematic Review

Atiya Rungjang, Jitlada Meephansan and Hok Bing Thio

Abstract

Paying attention to a microbial approach may lead to improvements in diagnosis, treatment, prevention, and prognosis of psoriasis. A systematic review was performed according to the PRISMA (Preferred Reporting Items for Systematic Reviews and Meta-Analyses) guidelines searching strategy to identify the pattern of the microbiome and the association of skin and gut microbiota with psoriasis, including the factors that may affect the results of the microbial study. In total, 16 studies were included in this systematic review. Ten studies investigated the skin microbiome, of which six studies were cross-sectional and four studies were prospective studies. Six studies investigated the gut microbiome, including five cross-sectional studies and one prospective study. The understanding of the relationship between microbiota and psoriasis may lead to diagnostics and treatment improvements. Currently, there is a slight consensus on some specific features that define psoriasis. However, no specific taxa have been identified as biomarkers of the disease, even from large-scale cohort studies. Thus, future cohort studies with standardized methodologies and proof-of-concept investigations in animal models may uncover the role of microbiota and the microbial pathways in psoriasis.

Keywords: psoriasis, microbiome, alpha diversity, beta diversity, dysbiosis

1. Introduction

Psoriasis is one of the most common immune-mediated inflammatory skin diseases. The prevalence of the disease has been reported, with ranges from 0.09 to 11.43% by the WHO Global Report 2016 [1]. Psoriatic skin lesions are characterized by hyperproliferation of keratinocytes, infiltration of immune cells, including neutrophils, T cells, dendritic cells, and macrophages. To date the etiology of this disease is not fully understood; genetic and environmental interaction plays a crucial role in the disease development [2, 3]. Recently, the immunological approach has helped to significantly clarify the pathophysiology of the disease. Dysregulation of both the bacteria, including *Staphylococcus aureus* [4], *Streptococcus pyogenes* [5], and fungi such as Malassezia [6] through innate and adaptive immune systems in genetically susceptible individuals, such as immune cells in the skin, Tumor necrosis factor α, dendritic cells—particularly pathogenic T cells that produce high levels of IL-17 in response to IL-23, all contribute substantially to the pathogenic process [7].

Previous studies have indicated an association between psoriasis and numerous comorbidities that share the chronic inflammatory state. Moreover, increasing evidence indicates that the gut microbe contributes to the onset of the low-grade inflammation, which is a pathological phenotype of these metabolic disorders [8].

Additionally, it has been known that several microorganisms contribute to psoriasis exacerbation alterations in the innate and adaptive immune processes [9]. The increasing evidence here suggests that the microbiota may play a critical role in psoriasis pathogenesis. This systematic review aims to elucidate the correlation between the microbiome and psoriasis pathogenesis, and the microbiota modulation that may lead to possible therapeutic interventions.

2. Psoriasis and microbiota

The initial search revealed a total of 629 studies of which 501 studies were excluded based on their title and abstract. The full texts were reviewed, and a further 116 studies were excluded. An additional four studies from the reference lists of already included studies were included in the systematic review. In total, 16 studies were included in this systematic review; 10 studies investigated the skin microbiome, of which 6 studies were cross-sectional and 4 studies were prospective study. Six studies investigated the gut microbiome, including five cross-sectional studies and one prospective study. The most commonly used method was 16S r RNA (skin swab, biopsies, curette); Langan et al. [10] used traditional culture combined with mass spectrometry (MALDI-TOF) (**Table 1**).

2.1 Skin microbiota in psoriasis

Several studies reported the characteristic features of microbiota in psoriatic skin (**Table 2**). Significant differences were observed between psoriatic lesion and control skin, but the changes were different in each study. Gao et al. [19] and Chang et al. [14] reported an increase in lesional skin diversity compared to non-lesional and control. In contrast, subsequent studies by Fahlen et al. [18] found wider range of Shannon index values in the control suggesting that the trend of decrease in lesional psoriasis microbiome diversity is consistent with the findings by Alexseyenko et al. [17] who observed a decrease in the diversity and significantly lower Shannon index in lesional skin. Consistent with previous studies, Tett et al. [16] found that psoriatic plaques at the ear are characterized by a significant decrease in microbial diversity. When beta diversity was analyzed to describe heterogeneity of microbial community, Fahlen et al. [18] reported a lower beta diversity in psoriasis compared to control, while Alexseyenko et al. [17] found that beta diversity was the highest in lesional skin, followed by unaffected skin, and the lowest in healthy skin. In line with the study by Tett et al. [16], which reported that ear lesions revealed higher beta diversity, Loeshe et al. [12] and Chang et al. [14] also reported a higher beta diversity at dry skin sites in psoriasis. At the phylum level, most skin bacterial composition fall into four major phyla: *Actinobacteria*, *Firmicutes*, *Bacteroidetes*, and *Proteobacteria*. Within these phyla, the three most abundant genera are: *Propionibacterium*, *Corynebacterium*, and *Staphylococcus*. From the studies of Gao et al. [19], Fahlen et al. [18], and Langan et al. [10], it has been revealed that at the phylum level, compared to healthy skin, psoriatic skin was associated with an increase in the relative abundance of *Firmicutes* but a decrease for *Actinobacteria*, which is partially consistent with Alexseyenko et al. [17] who identified psoriatic lesion as cutaneo type 2, which was dominated by *Firmicutes* and *Actinobacteria*. In contrast, *Firmicutes* were lower in the studies by Loeshe et al. [12], Assarsson et al. [13], and Drago et al. [15]. *Proteobacteria* showed inconsistent abundance, lower in lesional skin as observed by Gao et al. [19] whereas Fahlen et al. [18] observed an increase, and Drago et al. [15] reported that *Proteobacteria* and *Bacteroidetes* were the dominant microbiota in psoriasis lesion. At the genus

Study	Study design/ ratings of the quality	Population	Result
Langan et al. [10] (2019)	Cross-sectional/4	23 Pso, 20 C	Pso L At the phylum level: ↑ *Firmicutes*, ↓*Actinobacteria* At the genus level: ↑*Prevotella*, *Staphylococcus*, ↓*Anaerococcus* and *Propionibacterium* *Prevotella* and *Staphylococcus* significantly associated with Pso L Pso NL ↑*Anaerococcus*, *Propionibacterium*
	Prospective systemic treatment/2		*Actinobacteria*-to-*Firmicutes* ratio, partially reversible during treatment Biological therapies demonstrated the largest impact on the ratio of *Actinobacteria* to *Firmicutes* *Corynebacterium*, *Staphylococcus*, significantly correlated with PASI scores
Stehlikova et al. [11] (2019)	Cross-sectional/4	34 Pso, 25 C	Beta diversity: no significant differences between Pso L and Pso NL Pso L ↑ *Streptococcus* regardless of the sampling site ↑ *Brevibacterium* richness and evenness in in the elbow lesions, compared to back lesions ↓*Propionibacterium* PsoL, PsoNL compared C in elbow lesions Remark Alpha diversity and bacterial taxa from skin swab, scraping, and biopsy are comparable
Loesche et al. [12] (2018) Remark: study did not include healthy control	Cross-sectional/4	114 Pso	Beta diversity: Pso L > Pso NL Pso L At the phylum level: ↑*Actinobacteria* in leg, scalp, and trunk lesions ↓*Firmicutes* in scalp and trunk lesions At the genus level: ↓*Caulobacteraceae*, *Corynebacterium* leg lesions At the species level: ↑*Bacilli* ↓*Propionibacterium acnes* in scalp lesions *Streptococcus* colonization of skin does not correlate with severity in lesional and non-lesional skin
	Longitudinal RCT/1	89 Pso	Pso L and Pso NL respond similarly to ustekinumab Significant change in abundance from baseline in all body sites No difference diversity in Pso L vs. Pso NL except ↑ in trunk Pso L microbiota was not converging with Pso NL as treatment progressed Microbiota diverged further between Pso L and Pso NL across body sites
Assarsson et al. [13] (2018) Remark: study did not include healthy controls	Cross-sectional/4	26 Pso	Pso L ↓*Firmicutes Staphylococcus*
	Longitudinal Narrowband UVB/2		Pso L ↓*Firmicutes*, *Staphylococcus*, *Finegoldia*, *Anaerococcus*, *Peptoniphilus*, *Gardnerella*, *Prevotella*, *Clostridium* Pso NL ↓*Firmicutes* ↓*Pseudomonas* in treatment responders

Study	Study design/ ratings of the quality	Population	Result
Chang et al. [14] (2018)	Cross-sectional/4	28 Pso, 26 C	Alpha diversity: Pso L > Pso NL > C ↑ Beta diversity in all dry skin sites Psol L ↑ Alpha diversity at dry skin sites, with a trend at the sebaceous (scalp) site, and no increase at moist sites ↑*S. aureus* and *S. pettenkoferi* Pso NL ↑*S. sciuri* C ↑*P. acnes*, *P. granulosum*
Drago et al. [15] (2016)	Cross-sectional/4	3 adult first cousins— 1 AD, 1 Pso, 1 C (same lifestyle and environmental factors)	Pso L At the phylum level ↓ *Firmicutes*, ↑ *Proteobacteria* in Pso L compare to AD and C At the family level ↑ *Streptococcaceae*, *Rhodobacteraceae*, *Campylobacteraceae*, *Moraxellaceae* in Pso L compare to AD and C ↓*Staphylococcaceae*, *Propionibacteriaceae* in Pso L compare to AD, C. At the species level: ↓ *Propionibacterium* acnes in Pso compare to AD and C. ↓ *S. aureus* in Pso L < C < AD, no difference in Psp NL
Tett et al. [16] (2017) Remark: study did not include healthy control	Cross-sectional/4	28 Pso	Alpha diversity: Pso L < Pso NL in ear lesions (richness did not correlate with PASI score) Beta diversity: Pso L > Pso NL in ear lesions Pso L At the phylum level: *Actinobacteria* and *Firmicutes*. At the genus level: *Staphylococcus* At the species level: *S. epidermidis*, *P. acnes*, S. *caprae*/*capitis*, and *M. luteus*
Alekseyenko et al. [17] (2013)	Cross-sectional/4	75 Pso, 124 C	Alpha diversity: Pso L < Pso NL and C Beta diversity: Pso L > Pso NL > C. Pso L At the phylum level: Cutaneotype 2 (dominated by *Actinobacteria*, *Firmicutes*) At the genus level: ↑ combined relative abundance of *Corynebacterium*, *Propionibacterium*, *Staphylococcus*, *Streptococcus* ↓*Cupriavidus, Flavisolibacter*, *Methylobacterium*, *Schlegelella*. At the species level: *Acidobacteria*, *Schlegelella* *Acidobacteria* positively correlated with PASI C: Cutaneotype 1 (dominated by *Proteobacteria*) ↓ *Cupriavidus, Flavisolibacter*
	Longitudinal 12 weeks, 36 weeks after systemic treatment/1	17 Pso, 15 c	No statistically significant difference was observed between the lesion and unaffected groups, or longitudinally within groups Pso L ↑Relative abundance of *Corynebacterium*, *Propionibacterium*, *Staphylococcus* and *Streptococcus*

Study	Study design/ ratings of the quality	Population	Result
Fahlen et al. [18] (2012)	Cross-sectional/4	10 Pso, 12 C	Alpha diversity: no difference observed when using the Shannon index Beta diversity: Pso L < C Pso L At the phylum level: ↓ *Actinobacteria*, ↑ *Proteobacteria* in trunk lesions At the phylum level: ↓ *Propionibacterium* in all sites, ↓*Staphylococcus* ↑*Streptococcus/Propionibacterium* ratio
Gao et al. [19] (2008)	Cross-sectional/4	6 Pso, 6 C	Alpha diversity: Pso L > Pso NL, C Beta diversity: Pso L > C At the phylum level: ↑ *Firmicutes*, ↓ *Actinobacteria, Proteobacteria* At genus level: ↓ *Propionibacterium*, ↑ *Streptococcus* At the species level: ↓ *P. acne*, ↓ Anaerobic species

Pso—Psoriasis, C—Control, Pso L—Psoriasis lesional skin, Pso NL—Psoriasis non-lesional skin.

Table 1.
Skin microbiota in patients with psoriasis.

level, *Streptococcus* were higher in lesional skin by Gao et al. [19], Fahlen et al. [18], Alexseyenko et al. [17], Stehlikova et al. [11], and Drago et al. [15] while Loeshe et al. [12] found no correlation between psoriasis lesional and unaffected skin. *Staphylococcus* were detected more frequently in the lesion by Gao et al. [19] and Tett et al. [16] opposite to Fahlen et al. [18] who found that *Staphylococcus* were increased in abundance in healthy controls. Lower abundance of *Propionibacterium* in lesional skin was reported by Gao et al. [19], Fahlen et al. [18], Drago et al. [15], Stehlikova et al. [11], and Loeshe et al. [12], which is in contrast to Alexseyenko et al. [17] who reported an increase in the relative abundance of combined Gram positives such as *Corynebacterium*, *Propionibacterium*, *Staphylococcus*, and *Streptococcus*. In the subsequent study by Langan et al. [10], the presence of *Corynebacterium* and *Staphylococcus* was found to be significantly correlated with PASI scores while *Anaerococcus* and *Propionibacterium* were associated with non-lesional skin. These are consistent with the reports by Gao et al. [19] and Chang et al. [14] that at species level lesional skin psoriasis had an increased level of *S. aureus* but a decreased level of *P. acne*. On the other hand, study on the importance of site-specific microbiota without related disease reported that at the species level the most abundant bacteria were *S. epidermidis* and *P. acne* irrespective of disease status and hence suggested that an underlying subject-specific microbial signature better defines the microbiome.

There is a challenge to identify the explicit features of healthy or psoriasis microbiomes. Investigations of such a complex system of bacteria, fungi, and viruses are difficult and there is also high variation between samples. The composition of these communities of microorganisms depends on skin characteristics, such as sebaceous gland concentration, moisture content, topography, and temperature, as well as on host genetics and exogenous environmental factors [20]. Thus, the skin microbiome is biogeographically specific for each body site [21]. Demographic differences, such as gender, age, place of residence, living with animals, hygiene habits, occupation, and ethnicity also influence the composition of the skin microbiome [22]. The underlying disease and/or disease severity may also have an effect on the microbiome diversity or alterations in microbial communities due to disease states.

	Finding	By
Alpha diversity	Increased	Gao et al. [19], Chang et al. [14]
	Decreased	Fahlen et al. [18], Alexseyenko et al. [17], Tett et al. [16]
Beta diversity	Lower	Fahlen et al. [18]
	Highe	Alexseyenko et al. [17], Tett et al. [16], Loeshe et al. [12], Chang et al. [14]
Phylum level		
Firmicutes	Increased *Firmicutes*	Gao et al. [19], Fahlen et al. [18], Langan et al. [10]
	Decreased *Firmicutes*	Loeshe et al. [12], Assarsson et al. [13], Drago et al. [15]
Actinobacteria	Decreased *Actinobacteria*	Gao et al. [19], Fahlen et al. [18], Langan et al. [10]
	Increased *Firmicutes* and *Actinobacteria*	Alexseyenko et al. [17]
Proteobacteria	Decreased *Proteobacteria*	Gao et al. [19]
	Increased *Proteobacteria*	Fahlen et al. [18], Drago et al. [15]
Genus level		
Streptococcus	Increased *Streptococcus*	Gao et al. [19], Fahlen et al. [18], Alexseyenko et al. [17], Stehlikova et al. [11], Drago et al. [15]
	No correlation	Loeshe et al. [12]
Staphylococcus	Increased *Staphylococcus*	Gao et al. [19], Tett et al. [16]
	Decreased *Staphylococcus*	Fahlen et al. [18]
Propionibacterium	Lower *Propionibacterium*	Gao et al. [19], Fahlen et al. [18], Drago et al. [15], Stehlikova et al. [11], Loeshe et al. [12]
Gram positives	Increased relative abundance of combined Gram positives: *Corynebacterium*, *Propionibacterium*, *Staphylococcus*, and *Streptococcus*	Alexseyenko et al. [17]
	Corynebacterium and *Staphylococcus* were significantly correlated with PASI scores	Langan et al. [10]
	Increased *S. aureus,* decreased *P. acne*	Gao et al. [19]
	Site-specific microbiota without related disease	Tett et al. [16]

Table 2.
Summary of skin microbiota findings in psoriasis.

Moreover, microbiome diversities differ between studies; however, the more recent studies demonstrate decreased alpha diversity with increased beta diversity in psoriasis. Also, there are data that demonstrate a trend toward a changing microbial composition in psoriasis-affected skin. *Propionibacterium* is known as a protective commensal bacterium that is related with SCFA and propionate production, which regulates immune function. The decrease in the relative abundance of this microorganism in psoriasis may be related to the course of disease. In most studies, *Staphylococcus* are dominant in psoriatic skin, as species such as *S. aureus* proposed pathogenic Th17 activation while S. *epidermidis* appear to modulate immune and barrier functions. Interestingly, a study by Tett et al. [16] reported

S. epidermidis strains contain known virulence-related genes that are predominate in psoriasis-affected skin. Therefore, a future study at the species and the strain level may provide more information.

2.2 Gut microbiota in psoriasis

The fecal sample study by Scher et al. [23] revealed that gut microbiome in skin psoriasis and psoriatic arthritis had a decrease in alpha diversity compared to control, and *Actinobacteria* had a decrease in relative abundance at the phylum level. This is in line with Masallat et al. [24] who found that the relative abundance of *Actinobacteria* was reduced in psoriasis versus healthy controls with a negative correlation of PASI score whereas the ratio of *Firmicutes* to *Bacteroidetes* was positively correlated with PASI score. This is consistent with Codoner et al. [25] who found a lower abundance of *Bacteroides* at the genus level and characterized core microbiome of psoriasis by an increase in *Feacalibacterium* but a decrease in *Bacteroides* spp. The abundance of *Akkermansia*, *Ruminococcus*, and *Pseudobutyrivibrio* was found to be lower in psoriatic arthritis compared to controls by Scher et al. [23]. Eppinga et al. [26] found that the abundance of *Faecalibacterium prausnitzii* was reduced in psoriasis with a significant increase in the relative abundance of *Escherichia coli* (**Table 3**).

The gut is considered as a major immune organ, with gut-associated lymphoid tissue (GALT) being the most complex immune compartment [30]. It is well known that change in the microbe composition may promote both health and disease [31]. Intestinal dysbiosis has been implicated in the etiology of various diseases [32], such as Crohn's disease and obesity [33, 34]. Moreover, there is strong evidence that indicates intestinal dysbiosis is clinically relevant to psoriasis [35, 36]. The importance of the gut-skin axis in the pathogenesis of psoriasis has recently been documented in humans, as well as in animal models of psoriasis [9, 37]. A study by Tan et al. identified that the signature of gut microbiota and its function are significantly altered in the gut of patients with psoriasis [28]. Intestinal and skin microbiota directly regulate imiquimod-induced skin inflammation (IISI), and emphasizes the importance of microbiota in the pathogenesis of psoriasis [38]. A study by Zákostelská et al. has shown that exposure of mice to antibiotics inhibited the induction of psoriasis [37].

To identify bacterial pathways, which may be involved in the pathogenesis of psoriasis, it should be highlighted that SCFAs potentially regulate the generation and function of Th17 cells [39]. Moreover, in psoriasis the loss or depletion of *Faecalibacterium prausnitzii*, a major source of the protective SCFAs in the gut, may be associated with disease development [26]. In psoriatic arthritis, decreased *Akkermansia* and *Ruminococcus*, which are protective bacteria that regulate the intestinal barrier that produces SCFA, may be related with disease severity. Gut dysbiosis markedly reduced butyrate production, which inhibits NF-κβ, an inflammation pathway that impacts gut epithelial integrity and consequential cross-talk between gut proteins, bacteria, and the innate and humoral immune systems [27]. Alterations in the pathways involved in LPS function were also observed in psoriasis patients. Additionally, LPS is also thought to be involved in gut inflammation and has been linked to the pathogenesis of insulin resistance and diabetes mellitus [40], which has been epidemiologically associated with psoriasis. A decrease in *Bacteroides*, which are known to play an immunomodulatory role in the gut through the production of polysaccharide A that induces regulatory T cells, may result in an altered immune response [41]. Whereas a decrease in *Actinobacteria*, a phylum that includes the Bifidobacterium species that have been shown to reduce intestinal inflammation, suppresses autoimmunity, and induces regulatory T cell expression. There are also several studies that have shown how bacterial translocation from the

Study	Study design/ ratings of the quality	Population	Result
Shapiro et al. [27] (2019)	Cross-sectional/4	24 Pso 22 C	Alpha diversity, beta diversity: no significant differences At the phylum level: ↑*Firmicutes, Actinobacteria* ↓*Bacteroidetes, Proteobacteria* At the species level: ↑*Ruminococcus gnavus, Doreaformici generans* and *Collinsella aerofaciens, Prevotella copri* and *Parabacteroides distasonis*
Tan et al. [28] (2018)	Cross-sectional/4	14 Pso 14C	Pso L At the phylum level: ↓*Verrucomicrobia, Tenericutes* At the class level: ↓*Mollicutes, Verrucomicrobiae* At the order level: ↓*Verrucomicrobiales*, RF39 At the family level: ↓*Verrucomicrobiaceae*, S24–7 At the genus level: ↑*Bacteroidaceae*, *Enterococcaceae*, ↓*Akkermansia* At the species level: ↓*Akkermansia muciniphila*, ↑*Clostridium citroniae*
Codoner et al. [25] (2018) Remark: study did not include healthy control	Cross-sectional/4	52Pso compared with a cohort of over 300 healthy individuals extracted from the human microbiome project	Pso: ↑Beta diversity Enterotype 2 (predominance of *Prevotella*) tended to experience more frequent bacterial translocation and higher inflammatory status ↓*Bacteroides*, ↑*Akkermansia*, *Faecalibacterium*
Chen et al. [29] (2018)	Cross-sectional/4	32Pso 64 C	Diversity: no significant difference between Pso and C At the phylum level: ↑*Firmicutes*, ↓*Bacteroidetes* At the genus level: ↑ *Ruminococcus, Megasphaera* At the family level: ↓*Bacteroidaceae, Prevotellaceae* ↑ *Ruminococcaceae, Lachnospiraceae* Other covariates: sex, disease activity assessed by PASI score, phototherapy, arthritis, as well as diet, alcohol, smoking, coffee, tea, and habit of exercise, did not significantly affect the abundance profile of intestinal microbiota among Pso and C
	Patients receiving systemic treatment (DMARDs or biologics drugs BioSysDrug) subgroup analyses /2	20 Pso	↓the species *Prevotella stercorea*, belonging to *Prevotellaceae*, of the phylum *Bacteroidetes*, in patients receiving BioSysDrug

Study	Study design/ ratings of the quality	Population	Result
Scher et al. [23] (2015)	Cross-sectional/4	15Pso, 16PsA 17C	Diversity: Pso, PsA < C At the phylum level: ↓*Firmicutes, Clostridiales, Verrucomicrobiales* ↑ *Bacteroidetes in* PsA vs. Pso ↓*Actinobacteria* in Pso vs. C At the genus level: ↓*Coprococcus* spp. in Pso and PsA vs. C ↓*Akkermansia, Ruminococcus, Pseudobutyrivibrio* in PsA vs. C ↓*Parabacteroides, Coprobacillus* in Pso vs. C
Masallat et al. [24] (2016)	Case control/4	45Pso 45C	↑*Firmicutes/Bacteroidetes* ratio correlated with PASI ↓*Actinobacteria*
Pso—Psoriasis, C—Control, Pso L—Psoriasis lesional skin, Pso NL—Psoriasis non-lesional skin.			

Table 3.
Gut microbiota in patients with psoriasis.

gut and skin into the bloodstream may take place in psoriasis, and be responsible for driving the chronic, systemic inflammatory nature of the disease [42].

2.3 Skin mycobiota

An investigation by Findley et al. suggests that fungal diversity is increased in psoriatic lesions, compared to healthy skin sites. Furthermore the skin of psoriatic patients, at the genus level, is dominant with Malassezia [43]. Whereas the study by Takemoto et al. found that psoriatic skin revealed higher diversity and decreased relative abundance of Malassezia, which is still the most abundant phylum compared to controls. Moreover, the ratio of *M. globosa* to *M. restricta* is lower in psoriatic lesions [44]. Stehlikova et al. [11] found no significant difference in alpha and beta diversity and a significant increase in abundance of *M. restricta* in back lesions and *M. sympodialis* in the elbow lesions. Conversely, however, Paulino et al. showed that psoriatic lesions on the back, in decreasing order of abundance, are predominated by *M. restricta*, followed by *M. globosa* and *M. sympodialis*, respectively. Paulino et al. concluded that there was no significant difference between the fungal compositions of psoriatic and healthy skin [45]. Furthermore, Paulino et al. also showed there was no consistent variation between psoriasis and healthy controls [46] as *M. furfur* was found only in the skin of psoriasis participants in the study by Jagielski et al. [47] compared to healthy controls and atopic dermatitis.

2.4 Factors affecting microbiota study

So far, no specific patterns of microbiota in psoriatic patients have been identified (**Tables 1** and **3**).

The difficulty to establish such precise features, although a plethora of published studies have attempted to do so, is due to the lack of standardized protocols. Differences in sample collection and processing, sequencing methods, and analysis procedures between studies may impact the study results [48], and can confound comparisons and results in incompatible outcomes (**Table 4**).

	Findings	By
Alpha diversity	Decreased	Scher et al. [23]
Actinobacteria	Decreased	Scher et al. [23], Masallat et al. [24]
Bacteroides	Lower	Codoner et al. [25], Masallat et al. [24]
Feacalibacterium	Increased	Codoner et al. [25]

Table 4.
Summary of gut microbiota findings in psoriasis.

Different factors that may affect microbiome study

- Host factors: Not many studies accounted the interpersonal and intrapersonal factors that affect the microbial community.

- Samples collected from different body sites cannot be compared due to site-specific niches, as described previously [21].

- Sampling method: Several studies showed that different skin layers contain different bacterial communities [48]. Most of the published research on cutaneous microbiota has been based on skin swabs, which represent the surface of the skin. Prast-Nielsen et al. [49] found differences in both diversity and taxonomic composition of the microbiome obtained from swabs and biopsies of the same individual, while an investigation by Stehlikova et al. showed that various sampling approaches (swab, scraping, and biopsy) in affected and unaffected skin of psoriatic patients and in healthy control skin results in similar bacterial diversity despite the different genera abundance that is observed [11]. Grice et al. used three different sampling strategies in the antecubital fossa of five patients: swabs, skin scrapes, and punch biopsies, and concluded that similar microbial populations were captured by each technique and that the dominant species was present in the noninvasive swabs [50]. Recent studies have also reported that the tape stripping method may capture more viable bacteria than the swabbing method [51].

Sequencing methods, analysis procedures, and techniques for studying the microbiome:

- Langan et al. demonstrated that the changes in the microbiome during treatment that were detected by 16S rRNA were not detected by culture data. This suggested that changes in bacterial populations may have been too subtle to be detected by culture, or that changes are predominantly in nonculturable species [10].

- Studies using the most often used 16S rRNA have shown that the accuracy of molecular signatures depends on DNA sequencing and downstream analysis protocols. Therefore numerous combinations of primer pairs have been previously tested to select the most appropriate one for skin microbiome surveys; however, standardized methodology is still lacking [52].

    - Several studies suggested that primers for V1V3 and V3V4 hypervariable regions were described to sufficiently cover the skin bacterial diversity [20].

- ○ Statnikov et al. concluded that using 16S rRNA data from the V3–V5 locus leads to accurate and statistically significant molecular signatures, whereas data from the V1–V3 locus carry a limited diagnostic signal [53].

- ○ The latter study by Stehlikova et al. observed that variable regions of the V3V4 region capture a wider microbial diversity than the V1V2 region, where observed and estimated richness was significantly higher when using the V3V4 region compared to the V1V2 region [11].

- Whole-genome shotgun metagenomics offers the most comprehensive and robust data; however, as a result of its high cost, only a few shotgun metagenomic studies have been conducted on the microbiota associated with the skin, such as the Human Microbiome Project [54]. We found very few studies on psoriasis microbiome.

2.5 Therapeutic implements

Several studies reveal that psoriasis treatment changes the gut and skin microbiome, such as the correlation between psoriasis systemic treatment and the *Actinobacteria*-to-*Firmicutes* ratio. Biological therapies demonstrated the largest impact [10] during ustekinumab treatment; the composition of microbiota diverged further between lesional and non-lesional skin, across body sites, which could be due to the regression of lesions that returns the skin to more normal environments and increases the body site-specific niches [12]. Secukinumab (anti-IL17) therapy is associated with distinct and more profound gut microbiome shifts than ustekinumab therapy (anti-IL 12/23), in patients with psoriasis by increasing the relative abundance of *Proteobacteria* and decreases in *Bacteroidetes* and *Firmicutes* [55]. Burns et al. demonstrates that UVR has profound qualitative and quantitative influences on the composition of the skin microbiome by an increase in the phylum Cyanobacteria and a decrease in the family Lactobacillaceae and Pseudomonadaceae [56]. This suggests that skin microbiome alterations after UVB treatment could be related to treatment and treatment responses [13]. Thus, it may be implied that the modulation of the gut and skin microbiota can improve disease condition.

Therapeutic modalities that target the shifting microbiota:

The use of orally administered antibiotics, prebiotics, probiotics, and most recently, fecal transplantation [57] also appears to improve the disease condition and may be a practical prospect as a therapeutic avenue.

- Antibiotic treatment of psoriasis can alter the bowel flora toward normality, and therapy might include the use of appropriate antibiotics to reduce susceptible microbes while permitting others to flourish [58]. Saxena and Dogra reported that administration of benzathine penicillin in psoriasis vulgaris patients showed a significant improvement [59] and administration of azithromycin revealed a significant improvement at 12 weeks in the patient with psoriasis [60].

- Pro- and prebiotics are commonly used to modulate the microbiome by promoting the growth of specific species. Three studies using three distinct probiotic species affecting distinct pathways of the pathomechanism of psoriasis [61, 62] have all shown improvement in the course of the disease. The probiotics resulted in the improvement of epithelial barrier function, increased production of TNF-alpha by epithelial cells, and regulated activation of the NF-κβ pathway [63]. An issue with probiotic supplementation is that the colonization of probiotic bacteria in the gut is mostly transient as they are only detectable for less than 2 weeks after cessation of intake [64]. However, a study by Maldonado-Gómez et al. demonstrated that a certain *Bifidobacterium longum* (*B. longum*) strain was able to persist for over 6 months in a subset of subjects where it was originally absent [65]. A recent, randomized, double-blind, placebo-controlled trial evaluated the effect of a probiotic mixture as co-adjutant treatment together with topical steroids in 90 patients with plaque psoriasis. The results showed a large reduction in the score of severity indexes in the probiotic group, compared with the placebo group. Gut microbiota analysis demonstrated the efficacy of the probiotic in modulation of the composition of the microbiota. After the end of the probiotic or placebo intake, patients were followed up for 6 months. The results showed a lower risk of relapse in patients in the probiotic group [66].

- Topical probiotics show that after sequential applications of a donor microbiome, the recipient microbiome becomes more similar to the donor [67]. The use of topical probiotics may have special subclinical significance, for example, to improve skin defense with probiotic-containing cosmeceuticals. It has been reported that *B. longum* strains exert pro-differentiating, as well as and pro-regenerating, effects on primary human epidermal keratinocytes [68]. Thus, using the most suitable oral probiotic strain in combination with topical probiotics and/or prebiotics might help in the personalized treatment of skin disorders.

- Fecal microbiota transplantation (FMT) is currently being used to restore the balance of the intestinal microbiota [69, 70]. Particularly, this procedure has demonstrated >90% clinical resolution of recurrent, or refractory, *Clostridium difficile* infections [71]. Also, multiple FMTs seem to be able to induce remission in patients with IBD [72]. Due to these results, FMT is now being tested as a potential novel treatment for other gastrointestinal and extraintestinal diseases [73] as it greatly improves outcomes compared with those before treatment.

3. Conclusion

The function of microbiota may be more important in psoriasis. The metabolic activity of microbiota may become an upcoming research area in near future for identifying crucial biomarkers and new therapeutic approaches for psoriasis. Future cohort studies with standardized methodologies and proof-of-concept investigations in animal models may uncover the role of microbiota and the microbial pathway in psoriasis. This, then, may lead to the development of diagnostics and therapeutic opportunities.

Conflict of interest

The authors have no conflict of interest to declare.

Notes/Thanks/Other declarations

The authors gratefully acknowledge the Bualuang ASEAN Chair Professorship grant award provided by Thammasat University. The funders had no role in any parts of this study.

Author details

Atiya Rungjang[1], Jitlada Meephansan[1] and Hok Bing Thio[2*]

1 Division of Dermatology, Chulabhorn International College of Medicine, Thammasat University, Pathum Thani, Thailand

2 Department of Dermatology, Erasmus University Medical Center, Rotterdam, The Netherlands

*Address all correspondence to: h.thio@erasmusmc.nl

References

[1] WHO. Global Report on Psoriasis. World Health Organization; 2016. Available on the WHO website (www.who.\r\nint) or can be purchased from WHO Press. ISBN: 9789241565189. Available from: https://apps.who.int/iris/handle/10665/204417

[2] Vandenbroeck K. Cytokine Gene Polymorphisms in Multifactorial Conditions. Boca Raton/FL: CRC Press, Taylor & Francis Group; 2006. p. 459

[3] Mattozzi C, Richetta AG, Cantisani C, MacAluso L, Calvieri S. Psoriasis: New insight about pathogenesis, role of barrier organ integrity, NLR/CATERPILLER family genes and microbial flora. The Journal of Dermatology. 2012;**39**(9):752-760. DOI: 10.1111/j.1346-8138.2012.01606.x

[4] Tomi NS, Kränke B, Aberer E. Staphylococcal toxins in patients with psoriasis, atopic dermatitis, and erythroderma, and in healthy control subjects. Journal of the American Academy of Dermatology. 2005;**53**(1):67-72. DOI: 10.1016/j.jaad.2005.02.034

[5] Munz OH, Sela S, Baker BS, Griffiths CEM, Powles AV, Fry L. Evidence for the presence of bacteria in the blood of psoriasis patients. Archives of Dermatological Research. 2010;**302**:495-498. DOI: 10.1007/s00403-010-1065-0

[6] Rudramurthy SM, Honnavar P, Chakrabarti A, Dogra S, Singh P, Handa S. Association of *Malassezia* species with psoriatic lesions. Mycoses. 2014;**57**(8):483-488. DOI: 10.1111/myc.12186

[7] Griffiths CE, Barker JN. Pathogenesis and clinical features of psoriasis. Lancet. 2007;**370**(9583):263-271. DOI: 10.1016/S0140-6736(07)61128-3

[8] Correia B, Torres T. Obesity: A key component of psoriasis. Acta Biomedica. 2015;**86**(2):121-129

[9] Fry L, Baker BS, Powles AV, Fahlen A, Engstrand L. Is chronic plaque psoriasis triggered by microbiota in the skin? The British Journal of Dermatology. 2013;**169**(1):47-52. DOI: 10.1111/bjd.12322

[10] Langan EA, Künstner A, Miodovnik M, Zillikens D, Thaçi D, Baines JF, et al. Combined culture and metagenomic analyses reveal significant shifts in the composition of the cutaneous microbiome in psoriasis. British Journal of Dermatology. 2019. DOI: 10.1111/bjd.17989

[11] Stehlikova Z, Kostovcik M, Kostovcikova K, Kverka M, Juzlova K, Rob F, et al. Dysbiosis of skin microbiota in psoriatic patients: Co-occurrence of fungal and bacterial communities. Frontiers in Microbiology. 2019;**10**:438. DOI: 10.3389/fmicb.2019.00438

[12] Loesche MA, Farahi K, Capone K, Fakharzadeh S, Blauvelt A, Duffin KC, et al. Longitudinal study of the psoriasis-associated skin microbiome during therapy with ustekinumab in a randomized phase 3b clinical trial. Journal of Investigative Dermatology. 2018;**138**(9):1973-1981. DOI: 10.1016/j.jid.2018.03.1501

[13] Assarsson M, Duvetorp A, Dienus O, Söderman J, Seifert O. Significant changes in the skin microbiome in patients with chronic plaque psoriasis after treatment with narrowband ultraviolet B. Acta Dermato-Venereologica. 2018;**98**(4):428-436. DOI: 10.2340/00015555-2859

[14] Chang H-W, Yan D, Singh R, Liu J, Lu X, Ucmak D, et al. Alteration of the cutaneous microbiome in psoriasis and potential role in Th17 polarization. Microbiome. 2018;**6**(1):154. DOI: 10.1186/s40168-018-0533-1

[15] Drago L, De Grandi R, Altomare G, Pigatto P, Rossi O, Toscano M. Skin microbiota of first cousins affected by psoriasis and atopic dermatitis. Clinical and Molecular Allergy. 2016;**14**(1):2. DOI: 10.1186/s12948-016-0038-z

[16] Tett A, Pasolli E, Farina S, et al. Unexplored diversity and strain-level structure of the skin microbiome associated with psoriasis. NPJ Biofilms Microbiomes. 2017;**3**(1):14. DOI: 10.1038/s41522-017-0022-5

[17] Alekseyenko AV, Perez-Perez GI, De Souza A, et al. Community differentiation of the cutaneous microbiota in psoriasis. Microbiome. 2013;**1**(1):31. DOI: 10.1186/2049-2618-1-31

[18] Fahlén A, Engstrand L, Baker BS, Powles A, Fry L. Comparison of bacterial microbiota in skin biopsies from normal and psoriatic skin. Archives of Dermatological Research. 2012;**304**(1):15-22. DOI: 10.1007/s00403-011-1189-x

[19] Gao Z, Tseng C, Strober BE, Pei Z, Blaser MJ. Substantial alterations of the cutaneous bacterial biota in psoriatic lesions. PLoS One. 2008;**3**(7):e2719. DOI: 10.1371/journal.pone.0002719

[20] Grice EA, Segre JA. The skin microbiome. Nature Reviews. Microbiology. 2013;**9**(4):244-253. DOI: 10.1038/nrmicro2537

[21] Oh J, Byrd AL, Deming C, et al. Biogeography and individuality shape function in the human skin metagenome. Nature. 2014;**514**(7520):59-64. DOI: 10.1038/nature13786

[22] Fierer N, Hamady M, Lauber CL, Knight R. The influence of sex, handedness, and washing on the diversity of hand surface bacteria. Proceedings of the National Academy of Sciences of the United States of America. 2008;**105**(46):17994-17999. DOI: 10.1073/pnas.0807920105

[23] Scher JU, Ubeda C, Artacho A, et al. Decreased bacterial diversity characterizes the altered gut microbiota in patients with psoriatic arthritis, resembling dysbiosis in inflammatory bowel disease. Arthritis & Rhematology. 2015;**67**(1):128-139. DOI: 10.1002/art.38892

[24] Doaa M, Dalia M, Ahmed FS. Gut bacterial microbiota in psoriasis: A case control study. African Journal of. Microbiology Research. 2016;**10**(33):1337-1343. DOI: 10.5897/ajmr2016.8046

[25] Codoñer FM, Ramírez-Bosca A, Climent E, et al. Gut microbial composition in patients with psoriasis. Scientific Reports. 2018;**8**(1):3812. DOI: 10.1038/s41598-018-22125-y

[26] Eppinga H, Sperna Weiland CJ, Thio HB, et al. Similar depletion of protective *Faecalibacterium prausnitzii* in psoriasis and inflammatory bowel disease, but not in Hidradenitis suppurativa. Journal of Crohn's and Colitis. 2016;**10**(9):1067-1075. DOI: 10.1093/ecco-jcc/jjw070

[27] Shapiro J, Cohen NA, Shalev V, Uzan A, Koren O, Maharshak N. Psoriatic patients have a distinct structural and functional fecal microbiota compared with controls. The Journal of Dermatology. 2019;**46**(7):595-603. DOI: 10.1111/1346-8138.14933

[28] Tan L, Zhao S, Zhu W, et al. The *Akkermansia muciniphila* is a gut microbiota signature in psoriasis. Experimental Dermatology. 2018;**27**(2):144-149. DOI: 10.1111/exd.13463

[29] Chen Y, Ho HJ, Tseng C, Lai Z, Shieh J, Wu C. Intestinal microbiota profiling and predicted metabolic

dysregulation in psoriasis patients. Experimental Dermatology. 2018;**27**(12):1336-1343. DOI: 10.1111/exd.13786

[30] Ahluwalia B, Magnusson MK, Öhman L. Mucosal immune system of the gastrointestinal tract: Maintaining balance between the good and the bad. Scandinavian Journal of Gastroenterology. 2017;**52**(11):1185-1193. DOI: 10.1080/00365521.2017.1349173

[31] Blum HE. The human microbiome. Advances in Medical Sciences. 2017;**62**(2):414-420. DOI: 10.1016/j.advms.2017.04.005

[32] de Faria GF, Oliveira DG, de Oliveira JM, de Castro Ferreira LEVV, Cesar DE, Moreira APB. Influence of gut microbiota on the development and progression of nonalcoholic steatohepatitis. European Journal of Nutrition. 2018;**57**(3):861-876. DOI: 10.1007/s00394-017-1524-x

[33] Miele L, Giorgio V, Alberelli MA, De Candia E, Gasbarrini A, Grieco A. Impact of gut microbiota on obesity, diabetes, and cardiovascular disease risk. Current Cardiology Reports. 2015;**17**(12):120. DOI: 10.1007/s11886-015-0671-z

[34] Manichanh C, Rigottier-Gois L, Bonnaud E, et al. Reduced diversity of faecal microbiota in Crohn's disease revealed by a metagenomic approach. Gut. 2006;**55**(2):205-211. DOI: 10.1136/gut.2005.073817

[35] Park HS, Koh SJ, Park GY, et al. Psoriasis concurrent with inflammatory bowel disease. Journal of the European Academy of Dermatology and Venereology. 2014;**28**(11):1436-1441. DOI: 10.1111/jdv.12305

[36] Egeberg A, Mallbris L, Warren RB, et al. Association between psoriasis and inflammatory bowel disease: A Danish nationwide cohort study. The British Journal of Dermatology. 2016;**175**(3):487-492. DOI: 10.1111/bjd.14528

[37] Zákostelská Z, Málková J, Klimešová K, et al. Intestinal microbiota promotes psoriasis-like skin inflammation by enhancing Th17 response. PLoS One. 2016;**11**(7):e0159539. DOI: 10.1371/journal.pone.0159539

[38] Stehlikova Z, Kostovcikova K, Kverka M, et al. Crucial role of microbiota in experimental psoriasis revealed by a gnotobiotic mouse model. Frontiers in Microbiology. 2019;**10**(feb):236. DOI: 10.3389/fmicb.2019.00236

[39] Ohnmacht C, Park J-H, Cording S, et al. Mucosal Immunology. The microbiota regulates type 2 immunity through RORγt+ T cells. Science. 2015;**349**(6251):989-993. DOI: 10.1126/science.aac4263

[40] Cani PD, Amar J, Iglesias MA, et al. Metabolic endotoxemia initiates obesity and insulin resistance. Diabetes. 2007;**56**(7):1761-1772. DOI: 10.2337/db06-1491

[41] Mosca A, Leclerc M, Hugot JP. Gut microbiota diversity and human diseases: Should we reintroduce key predators in our ecosystem? Frontiers in Microbiology. 2016;7:455. DOI: 10.3389/fmicb.2016.00455

[42] Visser MJE, Kell DB, Pretorius E. Bacterial dysbiosis and translocation in psoriasis vulgaris. Frontiers in Cellular and Infection Microbiology. 2019;**9**:7. DOI: 10.3389/fcimb.2019.00007

[43] Findley K, Oh J, Yang J, et al. Human skin fungal diversity. Nature. 2013;**498**(7454):367-370. DOI: 10.1038/nature12171.Human

[44] Takemoto A, Cho O, Morohoshi Y, Sugita T, Muto M. Molecular characterization of the skin fungal microbiome in patients with psoriasis. The Journal of Dermatology. 2015;**42**(2):166-170. DOI: 10.1111/1346-8138.12739

[45] Paulino LC, Tseng CH, Strober BE, Blaser MJ. Molecular analysis of fungal microbiota in samples from healthy human skin and psoriatic lesions. Journal of Clinical Microbiology. 2006;**44**(8):2933-2941. DOI: 10.1128/JCM.00785-06

[46] Paulino LC, Tseng CH, Blaser MJ. Analysis of *Malassezia* microbiota in healthy superficial human skin and in psoriatic lesions by multiplex real-time PCR. FEMS Yeast Research. 2008;**8**(3):460-471. DOI: 10.1111/j.1567-1364.2008.00359.x

[47] Jagielski T, Rup E, Ziółkowska A, Roeske K, Macura AB, Bielecki J. Distribution of Malassezia species on the skin of patients with atopic dermatitis, psoriasis, and healthy volunteers assessed by conventional and molecular identification methods. BMC Dermatology. 2014;**14**(1):3. DOI: 10.1186/1471-5945-14-3

[48] Kong HH, Andersson B, Clavel T, et al. Performing skin microbiome research: A method to the madness. The Journal of Investigative Dermatology. 2017;**137**(3):561-568. DOI: 10.1016/j.jid.2016.10.033

[49] Prast-Nielsen S, Tobin A-M, Adamzik K, et al. Investigation of the skin microbiome: Swabs vs. biopsies. British Journal of Dermatology. 2019;**181**(3):572-579. DOI: 10.1111/bjd.17691

[50] Grice EA, Kong HH, Renaud G, et al. A diversity profile of the human skin microbiota. Genome Research. 2008;**18**(7):1043-1050. DOI: 10.1101/gr.075549.107

[51] Ogai K, Nagase S, Mukai K, et al. A comparison of techniques for collecting skin microbiome samples: Swabbing versus tape-stripping. Frontiers in Microbiology. 2018;**9**:2362. DOI: 10.3389/fmicb.2018.02362

[52] Meisel JS, Hannigan GD, Tyldsley AS, et al. Skin microbiome surveys are strongly influenced by experimental design. The Journal of Investigative Dermatology. May 2016;**136**(5):947-956. DOI: 10.1016/j.jid.2016.01.016

[53] Statnikov A, Alekseyenko AV, Li Z, et al. Microbiomic signatures of psoriasis: Feasibility and methodology comparison. Scientific Reports. 2013;**3**:1-7. DOI: 10.1038/srep02620

[54] Huttenhower C, Gevers D, Knight R, et al. Structure, function and diversity of the healthy human microbiome. Nature. 2012;**486**:207-214. DOI: 10.1038/nature11234

[55] Yeh N-L, Hsu C-Y, Tsai T-F, Chiu H-Y. Gut microbiome in psoriasis is perturbed differently during secukinumab and ustekinumab therapy and associated with response to treatment. Clinical Drug Investigation. 2019;**39**(12):1195-1203. DOI: 10.1007/s40261-019-00849-7

[56] Burns EM, Ahmed H, Isedeh PN, et al. Ultraviolet radiation, both UVA and UVB, influences the composition of the skin microbiome. Experimental Dermatology. 2019;**28**(2):136-141. DOI: 10.1111/exd.13854

[57] Mikó E, Vida A, Bai P. Translational aspects of the microbiome—To be exploited. Cell Biology and Toxicology. 2016;**32**:153-156. DOI: 10.1007/s10565-016-9320-6

[58] Wang GL, Li XY, Wang MY, et al. Cell-wall-deficient bacteria: A major etiological factor for psoriasis? Chinese Medical Journal.

2009;**122**(24):3011-3016. DOI: 10.3760/cma.j.issn.0366-6999.2009.24.019

[59] Saxena VN, Dogra J. Long-term use of penicillin for the treatment of chronic plaque psoriasis. European Journal of Dermatology. 2005;**15**(5):359-362. DOI: 10.1016/s0093-3619(08)70381-2

[60] Saxena VN, Dogra J. Long-term oral azithromycin in chronic plaque psoriasis: A controlled trial. European Journal of Dermatology. 2010;**20**(3):329-333. DOI: 10.1684/ejd.2010.0930

[61] Chen YH, Wu CS, Chao YH, et al. Lactobacillus pentosus GMNL-77 inhibits skin lesions in imiquimod-induced psoriasis-like mice. Journal of Food and Drug Analysis. 2017;**25**(3):559-566. DOI: 10.1016/j.jfda.2016.06.003

[62] Vijayashankar M, Raghunath N. Pustular psoriasis responding to probiotics—A new insight. Our Dermatology Online. 2012;**3**(4):326-328. DOI: 10.7241/ourd.20124.71

[63] Parisi R, Symmons DPM, Griffiths CEM, Ashcroft DM. Global epidemiology of psoriasis: A systematic review of incidence and prevalence. The Journal of Investigative Dermatology. 2013;**133**(2):377-385. DOI: 10.1038/jid.2012.339

[64] Frese S, Hutkins RW, Walter J. Comparison of the colonization ability of autochthonous and allochthonous strains of Lactobacilli in the human gastrointestinal tract. Advances in Microbiology. 2012;**2**(3):399-409. DOI: 10.4236/aim.2012.23051

[65] Maldonado-Gómez MX, Martínez I, Bottacini F, et al. Stable engraftment of *Bifidobacterium longum* AH1206 in the human gut depends on individualized features of the resident microbiome. Cell Host & Microbe. 2016;**20**(4):515-526. DOI: 10.1016/j.chom.2016.09.001

[66] Navarro-López V, Martínez-Andrés A, Ramírez-Boscá A, et al. Efficacy and safety of oral administration of a mixture of probiotic strains in patients with psoriasis: A randomized clinical trial. Acta Dermato-Venereologica. 2019;**9**:0. DOI: 10.2340/00015555-3305

[67] Paetzold B, Willis JR, Pereira de Lima J, et al. Skin microbiome modulation induced by probiotic solutions. Microbiome. 2019;**7**(1):95. DOI: 10.1186/s40168-019-0709-3

[68] Szöllősi AG, Gueniche A, Jammayrac O, et al. *Bifidobacterium longum* extract exerts pro-differentiating effects on human epidermal keratinocytes, in vitro. Experimental Dermatology. 2017;**26**(1):92-94. DOI: 10.1111/exd.13130

[69] Kragsnaes MS, Kjeldsen J, Horn HC, et al. Efficacy and safety of faecal microbiota transplantation in patients with psoriatic arthritis: Protocol for a 6-month, double-blind, randomised, placebo-controlled trial. BMJ Open. 2018;**8**(4):e019231. DOI: 10.1136/bmjopen-2017-019231

[70] Cammarota G, Pecere S, Ianiro G, Masucci L, Currò D. Principles of DNA-based gut microbiota assessment and therapeutic efficacy of fecal microbiota transplantation in gastrointestinal diseases. Digestive Diseases. 2016;**34**(3):279-285. DOI: 10.1159/000443362

[71] Van Nood E, Vrieze A, Nieuwdorp M, et al. Duodenal infusion of donor feces for recurrent *Clostridium difficile*. The New England Journal of Medicine. 2013;**368**(5):407-415. DOI: 10.1056/NEJMoa1205037

[72] Paramsothy S, Kamm MA, Kaakoush NO, et al. Multidonor intensive faecal microbiota transplantation for active

ulcerative colitis: A randomised placebo-controlled trial. Lancet. 2017;**389**(10075):1218-1228. DOI: 10.1016/S0140-6736(17)30182-4

[73] Cui B, Feng Q, Wang H, et al. Fecal microbiota transplantation through mid-gut for refractory Crohn's disease: Safety, feasibility, and efficacy trial results. Journal of Gastroenterology and Hepatology. 2015;**30**(1):51-58. DOI: 10.1111/jgh.12727

Chapter 5

Contribution of Gut Microbiome to Human Health and the Metabolism or Toxicity of Drugs and Natural Products

Prasat Kittakoop

Abstract

Trillions of microorganisms with a complex and diverse community are in the human gastrointestinal tract. Gut microbial genomes have much more genes than human genome, thus having a variety of enzymes for many metabolic activities; therefore, gut microbiota is recognized as an "organ" that has essential functions to human health. There are interactions between host and gut microbiome, and there are correlations between gut microbiome in the healthy state and in certain disease states, such as cancer, liver diseases, diabetes, and obesity. Gut microbiota can produce metabolites from nutrients of dietary sources and from drug metabolisms; these metabolites, for example, short-chain fatty acids (SCFAs), have substantial effects on human health. Drug-microbiome interactions play a crucial role in therapeutic efficiency. Some drugs are able to change compositions of gut microbiota, which can lead to either enhance or reduce therapeutic efficiency. This chapter provides an overview of roles of gut microbiota in human health and diseases and recent research studies on the metabolism or toxicity of drugs and natural products. Since gut bacteria considerably contribute to drug metabolism, research on the influence of gut microbiome on drug candidates (or natural products) should be part of the drug development processes.

Keywords: gut microbiota, gut microbiome, drug-microbiome interactions, drug-microbiota interactions, natural products-microbiome interactions, drug metabolism, drug toxicity, biotransformation, bioconversion

1. Introduction

The human gastrointestinal tract has various microorganisms, and "gut microbiota" has received attentions recently because the microbe population living in human intestine has significant effects to human health. Gut microbiota plays important roles in human, involving in many activities in a host body, for example, metabolism of xenobiotic compounds, immune system, nutrition, inflammation, and behavior. The delivery of prebiotics and probiotics to the human gastrointestinal tract, via dietary products or supplements, is one of the tools for management of microbiota in order to improve host health [1]. Moreover, gut microbiome has interactions with drugs and natural products, producing metabolites, which give effects on efficacy, metabolism, and toxicity of drugs. Gut microbiota plays a role in the metabolism of drugs and natural products, as well as nutrients in diet or food. The conversion of a dietary soybean isoflavone, daidzein (**1**) or genistein (**2**), to a bioactive compound, *S*-equol (**3**) (**Figure 1**) [2, 3], is a good example for the role of gut bacteria in the production of pharmacologically active agent in human because *S*-equol (**3**) is a potent ligand for estrogen receptor *β* [4]. Daidzein (**1**) is also derived from its corresponding isoflavone glycoside, daidzin (**4**), by *Bifidobacterium*, a representative of major bacterial species of human origin; this bacterium could transform daidzin (**4**) to daidzein (**1**) by cell-associated *β*-glucosidases (**Figure 1**) [5]. Moreover, *O*-desmethylangolensin (**5**) is also found as an intestinal bacterial metabolite of daidzein (**1**) [6, 7].

The transformation of achiral molecule daidzein (**1**) to a chiral molecule equol, which has one chiral center in its molecule, should provide two possible enantiomers of *S*-equol (**3**) and *R*-equol (**3R**) (**Figure 2**). However, gut bacteria selectively gives only *S*-equol (**3**), not *R*-equol (**3R**); this is interesting because only *S*-equol (**3**) has a high affinity to bind with estrogen receptor *β*, while *R*-equol (**3R**) has much less activity [4]. Therefore, *S*-equol (**3**), but not *R*-equol (**3R**), has high affinity for estrogen receptor *β* in human, and *S*-equol (**3**) has more potent estrogenic activity than estradiol [4]. In animal model, although a mixture of the two enantiomers of equol have the ability to inhibit bone loss in ovariectomized mice [8], *S*-equol (**3**) has better inhibitory effects on bone fragility than the racemic mixture containing both *S*-equol (**3**) and *R*-equol (**3R**) [9].

The ability of gut bacteria to selectively produce the correct bioactive isomer of *S*-equol (**3**) needed for human is intriguing. Shimada and co-workers identified enzymes involved in the bioconversion of daidzein (**1**) to *S*-equol (**3**) by the bacterium *Lactococcus* sp. strain 20–92, which was isolated from feces of healthy human [10]. The enzyme daidzein reductase catalyzes the transformation of daidzein (**1**) to (*R*)-dihydrodaidzein (**6**), which is in turn converted to (*S*)-dihydrodaidzein (**7**) by the enzyme dihydrodaidzein racemase (**Figure 2**). The enzyme dihydrodaidzein reductase catalyzes the conversion of (*S*)-dihydrodaidzein (**7**) to *trans*-tetrahydrodaidzein (**8**), which is converted to *S*-equol (**3**) by the enzyme tetrahydrodaidzein reductase [10]. The bioconversion of daidzein (**1**) selectively to *S*-equol (**3**), not *R*-equol (**3R**), by gut bacteria provides human the correct enantiomer for binding with estrogen receptor *β*; this may be host-bacterial mutualism in human intestine. An isoflavone daidzein (**1**) is found in leguminous plants such as soybeans and other

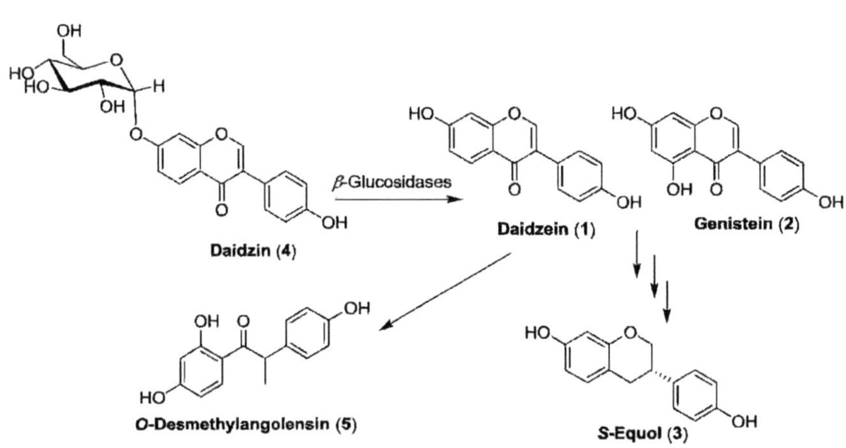

Figure 1.
Bioconversion of soybean isoflavones, daidzein (1), genistein (2), and daidzin (4), to S*-equol (3) and* O*-desmethylangolensin (5) by intestinal bacteria.*

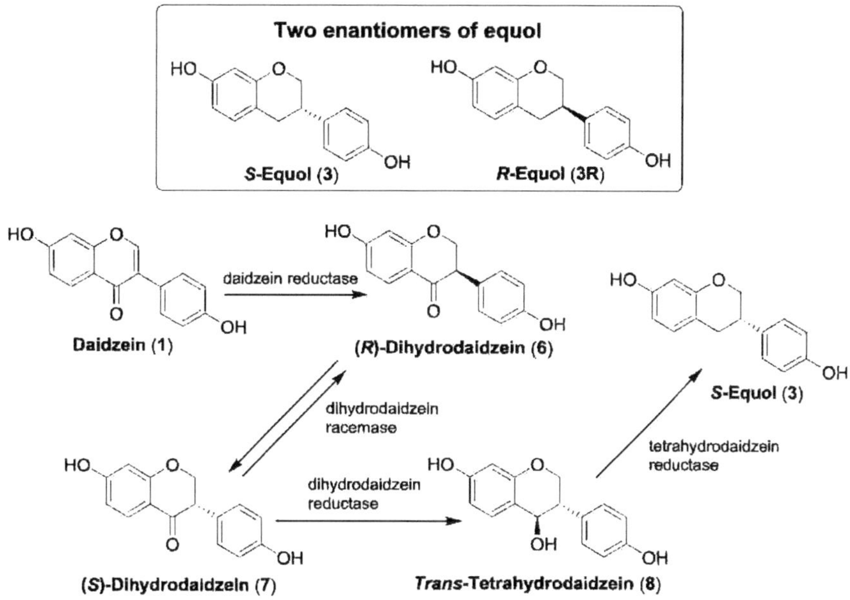

Figure 2.
Structures of two enantiomers of S-*equol (3) and* R-*equol (3R) and the bioconversion of daidzein (1) to* S-*equol (3) by the bacterium* Lactococcus *sp. through the metabolites* (R)-*dihydrodaidzein (6),* (S)-*dihydrodaidzein (7), and* trans-*tetrahydrodaidzein (8).*

legumes, which have been used as food for human since ancient times. Therefore, it is possible that gut bacteria have experienced with daidzein (**1**) long time ago, and their enzymatic evolutions lead to the selective bioconversion of daidzein (**1**) to *S*-equol (**3**), which has biological activity for human. Interestingly, many studies revealed that there is the intestinal microbiota-to-host relationship, i.e., a cross talk, between gut microbiota and human host and interactions between gene products from the microbiome with metabolic systems of human diseases such as obesity and diabetes [11].

The conversion of a dietary soybean isoflavone, daidzein (**1**) or genistein (**2**), to *S*-equol (**3**), by gut bacteria has been known for many years; however, scientists might not be aware of the importance of gut microorganisms in the past. Recently, a number of studies have revealed many essential roles of gut microbiota in human health and diseases. Gut microbiome can transform nutrients and dietary fibers to produce bioactive metabolites, for example, short-chain fatty acids (SCFAs) and nicotinamide, which have a significant impact on human health and diseases. There have been reports on interactions of gut microbiome and compounds, e.g., drugs and natural products, after humans take these compounds as drugs for the treatment of diseases. The metabolites obtained from the metabolism of drugs/natural products by the activities of gut microbiome have either positive or negative effects on therapeutic efficiency. This chapter provides the information of recent studies on the influence of the metabolites produced by gut microbiome on human health and diseases and on the interactions of microbiome and drugs/natural products.

2. Contributions of metabolites produced by gut microbiome to human health and diseases

The human gastrointestinal tract has trillions of microorganisms with a complex and diverse community. Gut microbiome is recognized as an “organ” because gut microorganisms have metabolic activities similar to an organ and have several essential functions to human health [12]. It is estimated that microbial cells in the human body are 10 times more than human cells and that gut microbiome has 150 times more genes than human genome [13]. Perturbation of gut microbial communities leads to the imbalance of gut microorganisms, by either reducing or increasing particular microbial species or altering the relative abundance of certain microorganisms; this is collectively known as "dysbiosis." Microbial dysbiosis can cause certain diseases such as irritable bowel syndrome, diabetes, cancer, inflammatory bowel diseases, and obesity [14–16]. Gut microorganisms are able to produce many metabolites, which give substantial contributions to human health because they are involved in various physiological processes, i.e., host immunity, cell-to-cell communication, and energy metabolism [17, 18]. The metabolites produced by gut microbiome are linked with human diseases, for example, colorectal cancer [19], depression [20], inflammation and cancer [21], and cardiovascular and metabolic diseases [22, 23]. Among the metabolites produced by gut microbiome, SCFAs considerably play critical roles in human health. Gut microbiome produces acetate (**9**), propionate (**10**), and butyrate (**11**) (**Figure 3**), the respective conjugate bases of acetic acid, propionic acid, and butyric acid; these SCFAs are from saccharolytic fermentation of dietary fibers by gut microorganisms [24]. Butyrate (**11**) from the metabolism of gut microbiome could induce differentiation of colonic regulatory T cells in mice, suggesting that gut microorganisms are substantially involved in immunological homeostasis in the gastrointestinal tract of human [25]. SCFAs produced by gut microbiota are significantly linked with hypertension and kidney diseases [26]. SCFAs are vital fuels for intestinal epithelial cells and can maintain intestinal homeostasis; they are involved in the regulation of gut epithelial cells and immunity that is relevant to inflammatory bowel diseases [27, 28]. SCFAs are able to activate G-protein-coupled receptor, for example, GPR43, which has a role in intestinal inflammatory diseases, i.e., inflammatory bowel diseases [29]. Moreover, SCFAs produced by gut microbiome are energy source for colonocytes and can inhibit histone deacetylases, the enzymes catalyzing the removal of acetyl groups from the lysine residue of histone [30]. Recent study revealed that butyrate (**11**) from the metabolism of gut microbiome could promote histone crotonylation in colon epithelial cells and that the reduction of the gut microbiota leads to many changes in histone crotonylation in the colon [31].

Recent study revealed that SCFAs produced by gut microbiome had relationships with metabolic diseases [32]. The level increase of butyrate (**11**) (**Figure 3**) by gut microbiome can improve insulin response after an oral glucose tolerance test; moreover, the defects in the production or absorption of propionate (**10**) led to an increased risk of type 2 diabetes [32]. Previous study also demonstrated that type 2 diabetes is linked with changes in the composition of gut microbiome because the profile of gut microorganisms in human with type 2 diabetes is different from that without type 2 diabetes (a control group) [33]. SCFAs are known to have a significant impact on the energy homeostasis, i.e., controlling the energy metabolism; therefore, modulation of SCFAs could be a nutritional target to prevent diseases associated with metabolism disorders, for example, type 2 diabetes and obesity [34]. Gut microbiome is also linked with food allergy in human, and changes in the

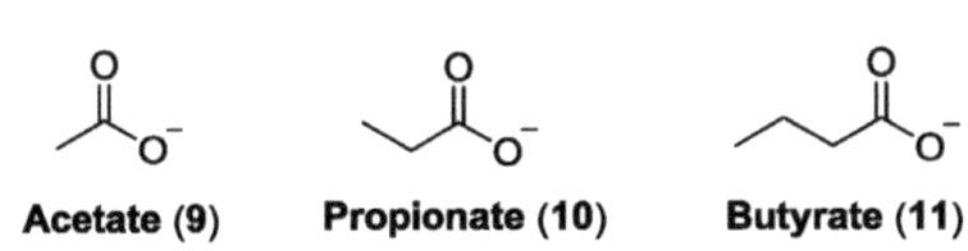

Figure 3.
Structures of SCFAs including acetate (9), propionate (10), and butyrate (11).

population and composition of gut microbiota might cause food allergy [35], and the use of gut microbiome is a potential innovative strategy to prevent food allergy in human [36]. In an animal model, certain gut bacteria, e.g., *Clostridia* species, might be useful for prevention or therapy of food allergy [37]. Recent investigation led by Nagler showed that butyrate (**11**) (**Figure 3**) produced by the gut bacterium, *Anaerostipes caccae*, could contribute to the prevention of milk allergy in children [38]. Germ-free mice colonized with bacteria in feces of healthy infants can protect mice against milk allergy, while those colonized with bacteria in feces of milk allergic infants could not protect mice from milk allergy; this result indicates that gut microbiotas are involved in milk allergy. Detailed analysis revealed that compositions of gut bacteria in healthy infants were different from milk allergic infants, and the gut bacterium, *A. caccae*, was the key agent to protect against an allergic response to food [38]. *A. caccae* is a saccharolytic intestinal bacterium producing butyrate (**11**) [39]. It is known that butyrate (**11**) is a key energy source for colonic epithelial cells, regulating energy metabolism and autophagy in the mammalian colon [40]. Therefore, butyrate (**11**) is likely to be the key metabolite responsible for the protection of milk allergy [38]. An independent study revealed that a dietary supplemented with the bacterium *Lactobacillus rhamnosus* could promote tolerance in infants with cow's milk allergy by enrichment of butyrate-producing bacterial strains [41]. The increased levels of butyrate (**11**) in feces of infants who received the supplement with *L. rhamnosus* were observed in the most tolerant infants against milk allergy [41].

Nicotinamide (**12**) is an amide derivative of vitamin B3 or niacin or nicotinic acid (**13**) (**Figure 4**) and is a substrate for nicotinamide adenine dinucleotide (NAD), a coenzyme in many important enzymatic oxidation–reduction reactions, for example, electron transport chain, citric acid cycle, and glycolysis. Nicotinamide (**12**) is known to have a role in neuronal systems in the central nervous system, thus implicating in neuronal death and neuroprotection [42]. Recent study led by Elinav revealed that nicotinamide (**12**) produced by the gut bacterium, *Akkermansia muciniphila*, significantly protected the progression of the neurodegenerative disease, amyotrophic lateral sclerosis (ALS) [43]. The experiment demonstrated that removal of gut microorganisms by treating mice with antibiotics could promote the ALS symptoms in mice, indicating that gut microbiome modulated the progress of ALS disease [43]. The study showed that the species of gut bacteria in healthy human were different from that in ALS patients; *A. muciniphila* was abundant in healthy people, while *Ruminococcus torques* and *Parabacteroides distasonis* were relatively abundant in ALS patients. Remarkably, transplantation of gut bacteria from human gut to germ-free mice revealed that the gut bacterium *A. muciniphila* improved the ALS symptoms, while the gut bacteria *R. torques* and *P. distasonis* worsened the ALS symptoms [43]. Detailed analysis found that the gut bacterium *A. muciniphila* provided nicotinamide (**12**) as a bioactive metabolite that improves the ALS symptoms. Indeed, a direct injection of nicotinamide (**12**) into mice with ALS could improve a motor-neuron function. The study in 37 patients

Figure 4.
Structures of nicotinamide (12) and niacin or nicotinic acid (13).

with ALS revealed that the levels of nicotinamide (**12**) in cerebrospinal fluid of ALS patients were lower than that in people without ALS. Moreover, analysis of microbial genes involved in nicotinamide synthesis in feces of ALS patients revealed that people with ALS had less number of the genes for nicotinamide synthesis; these genes were mainly from the gut bacterium *A. muciniphila*. Therefore, it is likely that ALS patients might have less abundance of *A. muciniphila* in their gastrointestinal tract [43]. This work suggests that gut microbiome has a significant link with human disease pathophysiology and that there is an opportunity to use microbial therapeutic targets for certain diseases. Indeed, a clinical trial on human using the gut bacterium, *A. muciniphila*, in overweight and obese insulin-resistant volunteers demonstrated that the gut bacterium could reduce insulin resistance indices and could lower the levels of circulating insulin and blood cholesterol, thus improving the profile of blood lipid and insulin sensitivity [44]. This microbial therapeutic approach is safe and may be applied for the treatment of overweight or obese insulin-resistant people.

It is known that gut microbiota is significantly associated with autism spectrum disorder, a form of mental disorder with difficulties in social communication and interaction [45]. Intriguingly, a recent study led by Sharon and Mazmanian revealed that gut microbiota could produce neuroactive metabolites which contribute to the pathophysiology of autism spectrum disorder, thus regulating behaviors in mice [46]. The experiment showed that germ-free mice receiving gut microbiota from human donors with autism spectrum disorder could induce autistic behaviors in mice. The metabolites produced by gut bacteria, 5-aminovaleric acid (**14**) and taurine (**15**) (**Figure 5**), were found to modulate behaviors related to autism spectrum disorder. Both 5-aminovaleric acid (**14**) and taurine (**15**) are $GABA_A$ receptor agonists [47, 48]. Levels of 5-aminovaleric acid (**14**) in mice with autism spectrum disorder were significantly lower than that in the control mice, while levels of taurine (**15**) in mice with autism spectrum disorder were ca. 50% less than the control group [46]. Administration of 5-aminovaleric acid (**14**) and taurine (**15**) to mice with autism spectrum disorder could improve repetitive and social behaviors, i.e., modulating neuronal excitability in mice brain and improving behavioral abnormalities [46]. This finding suggests that autism spectrum disorder is also related to the influence of gut microbiota; therefore microbiome interventions using fecal microbiota transplantation, as well as supplementation with metabolites produced by gut microorganisms or with probiotics, may improve the quality of life for people with autism spectrum disorder.

Gut microbiome substantially contributes to human health and diseases, and the metabolites produced by gut microbiome mentioned earlier underscore the importance of gut microorganisms in health and certain diseases in human. Health and diseases of individuals partly rely on the conditions of gut microbiome whether they have healthy gut microbiota or unhealthy ones. Gut microbiota is therefore considered as a "hidden" or "forgotten" human organ [12], involving in pathology of Alzheimer's disease [49], endocrine organ involving metabolic diseases [50], and

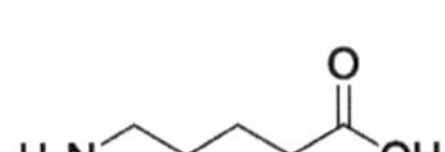

5-Aminovaleric acid or 5-Aminopentanoic acid (14)

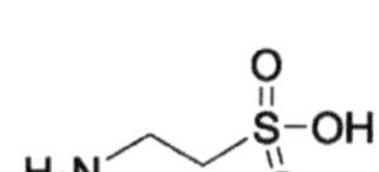

Taurine or 2-Aminoethanesulfonic acid (15)

Figure 5.
Structures of 5-aminovaleric acid or 5-aminopentanoic acid (14) and taurine or 2-aminoethanesulfonic acid (15).

chronic gastrointestinal disease [51]. Moreover, gut microbiota is also considered as an "invisible" organ that controls and manipulates the function of drugs [52]. The imbalance of gut microbiota, or known as dysbiosis, leads to unhealthy conditions for human or even causes certain diseases. Therefore, the use of gut microbiota as a therapeutic target for treatments of human diseases is an emerging approach for many diseases, for example, Parkinson's disease [53], cardiovascular disease [54], metabolic disorders [55], hepatocellular carcinoma [56], nonalcoholic fatty liver disease [57], food allergy [58], and heart failure [59]. Supplementation with probiotics or with health-promoting bacteria is a possible therapeutic method and may widely be used in the near future. Fecal microbiota transplantation or supplementation with metabolites from gut microorganism needs more clinical studies; the two approaches will be a challenging research on gut microbiota in the near future.

3. Interactions of gut microbiome and drugs and/or natural products

It is estimated that a total mass of bacteria in the human body is around 0.2 kg (for people with a weight of 70 kg) and that the densities of commensal microorganisms in the human gastrointestinal tract ranged from 10^8 to 10^{11} bacterial cells/g [60]. Oral administration of drugs delivers drugs to the gastrointestinal tract that contains high densities of gut microorganisms, which could encode 150-fold more genes than those of the human genome [61]; therefore, gut microbes are able to encode many enzymes with drug-metabolizing potential [62]. Gut microbiota is recognized as an "invisible organ" responsible for controlling drug functions and modulation of drug metabolism processes [52]. Normally, antibiotic drugs give direct effects toward microorganisms in the human gastrointestinal tract, providing either negative or positive (beneficial) effects to the composition of gut microbiota [63]. However, intestinal microbiota have many important roles in maintenance of human health; therefore, perturbation of gut microbiome by antibiotics could give negative impact to human, for example, loss of colonization resistance that can prevent invading microorganisms colonizing in the human gastrointestinal tract [64]. In addition to antibiotic drugs, a recent study led by Typas demonstrated that nonantibiotic drugs also gave extensive impact on human gut bacteria because around 24% of 1197 drugs showed antibacterial activity toward at least one strain of gut bacteria [65]. This is considered as "antibiotic-like side effects" of nonantibiotic drugs, which could potentially promote antibiotic resistance that is one of the major public health problems worldwide. This finding provides essential information for drug discovery research, i.e., addressing a potential new side effect of drugs and repurposing of nonantibiotic drugs as antibacterial agents.

The next sections will highlight the interactions of gut microbiome, especially the chemistry of the drug metabolites produced by gut microorganisms, toward certain drugs and natural products. The metabolism of drugs or natural products by gut microbiome could lead to the production of bioactive metabolites, which have either beneficial effects or negative properties (i.e., reducing efficacy of drugs or natural products). The study on the interactions of gut microbiota and drugs or natural products as part of drug development process is discussed in the next sections.

3.1 Interactions of gut microbiome and commonly used drugs

Once drugs enter the human gastrointestinal tract, they encounter trillions of microorganisms, which are able to encode 150-fold more genes than human genome [61]. A number of enzymes encoded by gut microbial genes catalyze the biotransformation of drugs, producing bioactive metabolites, which have effects on human health [60]. Advances in liquid chromatography-mass spectrometry (LC–MS) technology allow the identification of the metabolites produced by gut microbiome, as well as detailed study of pharmacokinetics of drugs and their metabolites, while genome sequencing substantially assists the identification of genes encoding enzymes in gut microorganisms. Zimmermann and co-workers investigated the drug metabolism of an antiviral nucleoside drug, brivudine (**16**), which is used for the treatment of herpes zoster virus; the study was performed using mice inoculated with mutant microbiota [66]. It was found that the bioconversion of brivudine (**16**) to bromovinyluracil (**18**) (or 5-(*E*)-(2-bromovinyl)uracil) was achieved by enzymes from both mammalian cells and gut microbial communities isolated from mice, suggesting that both host and microbiota are capable of such biotransformation (**Figure 6**). Previously, intestinal anaerobic bacteria were found to convert another antiviral drug, sorivudine (**17**), to bromovinyluracil (**18**) (**Figure 6**) [67].

Gut bacteria, *Bacteroides thetaiotaomicron* and *B. ovatus*, were the major species having the highest metabolic activity to convert brivudine (**16**) to bromovinyluracil (**18**) [66]. Comparison of serum kinetics of brivudine (**16**) and bromovinyluracil (**18**) in conventional (a control with bacteria) and germ-free mice after feeding with the drug brivudine (**16**) suggested that intestinal bacteria contributed to the amount of bromovinyluracil (**18**) in serum because the level of bromovinyluracil (**18**) in conventional mice serum was five times higher than that of germ-free mice [66]. The gene, *bt4554*, encoding the enzyme purine nucleoside phosphorylase necessary for the metabolism of brivudine (**16**) is present in *B. ovatus* and conserved in the bacterial phylum *Bacteroidetes*; the expression of the gene *bt4554* is a rate-limiting step [66]. The gut bacterium, *B. thetaiotaomicron*, was found to completely metabolize the drug brivudine (**16**) to bromovinyluracil (**18**), which is absorbed from both the cecum and colon. This study was also able to predict the levels in serum and sources of the metabolite bromovinyluracil (**18**) derived from a drug sorivudine (**17**) (**Figure 6**) [66].

Zimmermann and co-workers also used clonazepam (**19**) (**Figure 7**), an anticonvulsant and antianxiety drug, as a model [66]; the metabolism of this drug in rats gave metabolic products through nitroreduction, oxidation, glucuronidation, and enterohepatic cycling [68]. After an oral administration of a drug clonazepam (**19**) to mice, 7-NH_2-clonazepam (**20**) and 7-NH_2-3-OH-clonazepam (**21**) were found as major metabolites in serum of the conventional mice (**Figure 7**). The host-microbiome pharmacokinetic model revealed that 7-NH_2-clonazepam (**20**) in serum was substantially from a microbial contribution. Experiments also revealed that intestinal microbes could convert glucuronyl-3-OH-clonazepam (**23**) to 3-OH-clonazepam (**22**), which in turn transformed to 7-NH_2-3-OH-clonazepam (**21**) by microbial reduction (**Figure 7**) [66]. The study established a pharmacokinetic model that can predict microbiome or host (human) contributions to drug

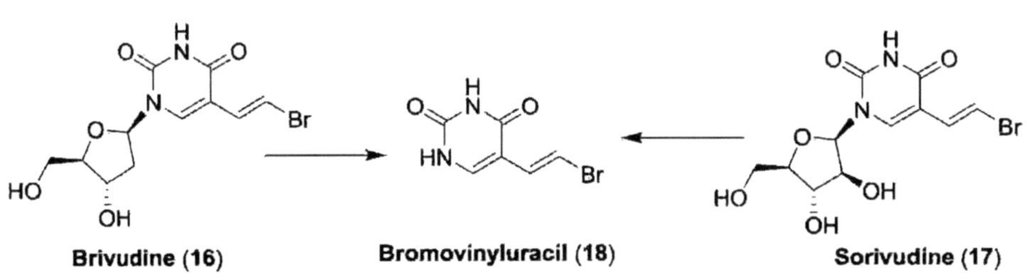

Figure 6.
Biotransformation of antiviral drugs brivudine (16) and sorivudine (17) to bromovinyluracil (18) by gut bacteria.

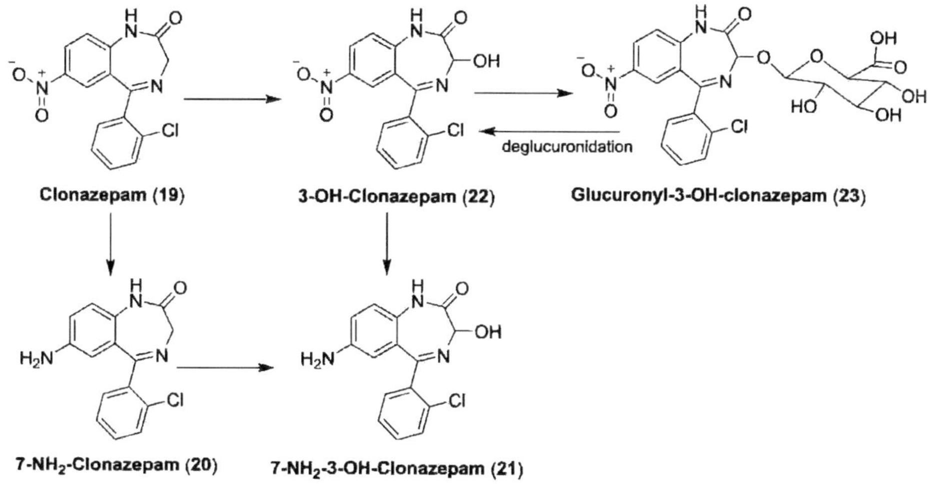

Figure 7.
Biotransformation of clonazepam (19) to the metabolites 20–23 by human intestinal microbes.

metabolism, e.g., the ability to distinguish drug-metabolizing activity by human or gut bacteria [66]. This research model is particularly useful for the study on drug metabolism in an animal model.

Gut microbiome has potential ability to metabolite many drugs, thus affecting the therapeutic efficacy due to lower concentrations of drugs. The study on the drug metabolism of 271 commonly used drugs by gut bacteria revealed that, after incubation of drugs with gut bacteria, the levels of 176 drugs (accounting for two thirds of 271 drugs) were significantly reduced, indicating that these drugs were metabolized by gut bacteria [62]. Intriguingly, each bacterial strain (from 76 human gut bacterial strains) could metabolize up to 11–95 drugs [62]. This result suggests that, when designing the drug molecules, the drug metabolism by gut microbes should be seriously considered, particularly the drugs delivered by an oral administration. Therefore, the action of gut microbiome toward individual drug candidates should also be evaluated during the drug development process. Untargeted metabolomics analysis is used to identify products derived from drug metabolism by gut bacteria, and it could properly identify the metabolites from microbial metabolism of drugs [62]. Some drugs, for example, paliperidone, sulfasalazine, and pantoprazole, were previously investigated for their metabolism by gut microbes [69]. Detailed analysis by high-resolution mass spectrometry (HRMS) revealed that drugs with an acetyl ester or an alkene functional group, such as norethisterone acetate (**24**), drospirenone (**25**), and roxatidine acetate (**26**), were metabolized through either deacetylation (removing C_2H_2O) or hydrogenation (adding H_2) by gut bacteria (**Figure 8**) [62]. Gut bacteria metabolized drugs with aliphatic hydroxyl or amine functional group such as dasatinib (**27**), fluphenazine (**28**), and primaquine (**29**) through propionylation (adding C_3H_4O), giving their corresponding *O*- or *N*-propionyl products **30**, **31**, and **32**, respectively (**Figure 8**). The HRMS data clearly indicated the mass difference of 56.026 unit of a propionyl group between the drug and its corresponding derivative [62].

Zimmermann and co-workers investigated the metabolism of drug in mice model and in human gut microbial communities using a corticosteroid drug, dexamethasone (**33**), as a model (**Figure 9**) [62]. It is known that this class of drug is metabolized by the bacterium *Clostridium scindens* through the side-chain cleavage, known as the desmolytic activity, to produce the active androgen form of the drug, dexamethasone-desmo (**34**) (**Figure 9**) [70, 71]. Levels of dexamethasone-desmo

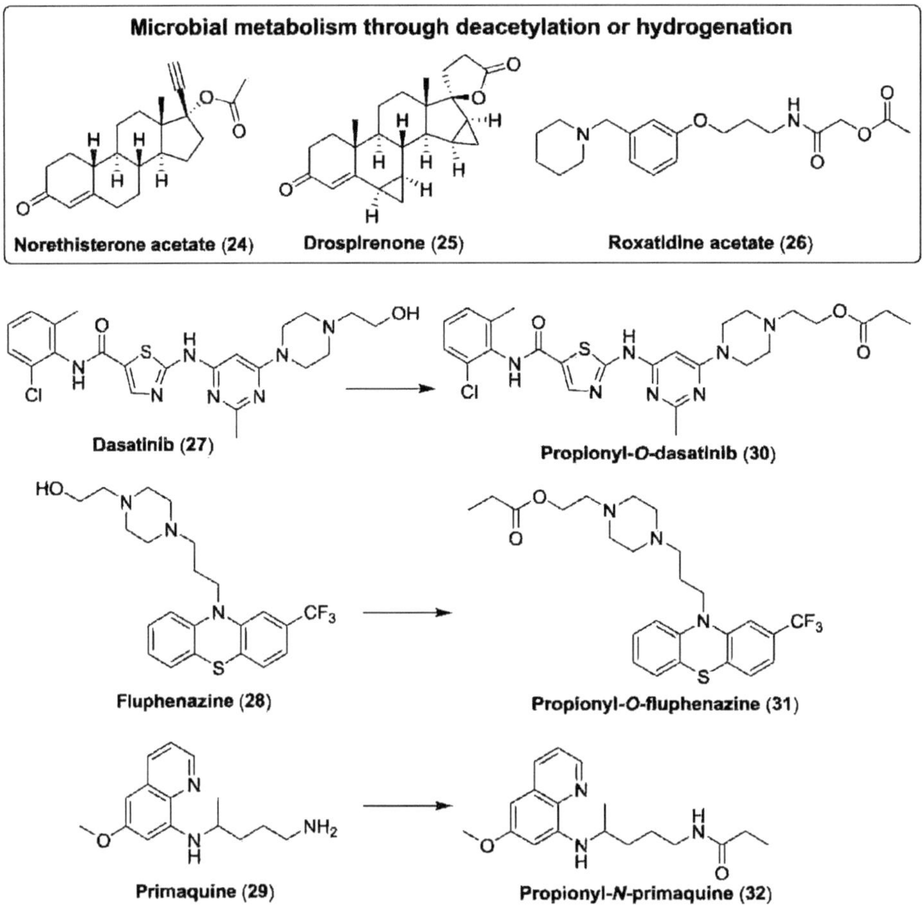

Figure 8.
Structures of the drugs, norethisterone acetate (24), drospirenone (25), and roxatidine acetate (26) and biotransformation of dasatinib (27), fluphenazine (28), and primaquine (29) to their corresponding products 30, 31, and 32, respectively, by gut bacteria.

(**34**) were measured after an oral administration of dexamethasone (**33**) to germ-free mice and to mice that have only one bacterial species of *C. scindens*, technically known as gnotobiotic mice ($GN^{C.\ scindens}$). Although dexamethasone (**33**) was found in the cecum of germ-free mice and gnotobiotic mice, the levels of the drug were significantly reduced in gnotobiotic mice, suggesting that the bacterium *C. scindens* associated in these mice is involved in the drug metabolism. Accordingly, levels of the androgen form of the drug, dexamethasone-desmo (**34**), which are derived from the metabolism of dexamethasone (**33**), were higher in both serum and cecum of gnotobiotic mice than that of germ-free mice [62]. Moreover, similar corticosteroid drugs, i.e., prednisone (**35**), prednisolone (**36**), cortisone (**37**), and cortisol (**38**), were also metabolized by the intestinal bacterium *C. scindens* through the desmolytic activity, giving the metabolite products of prednisone-desmo (**39**), prednisolone-desmo (**40**), cortisone-desmo (**41**), and cortisol-desmo (**42**), respectively (**Figure 9**). However, when incubating the drug dexamethasone (**33**) with gut bacterial community isolated from 28 healthy human participants under anaerobic condition, the drug-metabolizing activity had considerable interpersonal variation as suggested by level variations of the drug metabolite, dexamethasone-desmo (**34**) [62]. This result implies that dexamethasone (**33**) is also metabolized by other gut bacterial species, not only *C. scindens*.

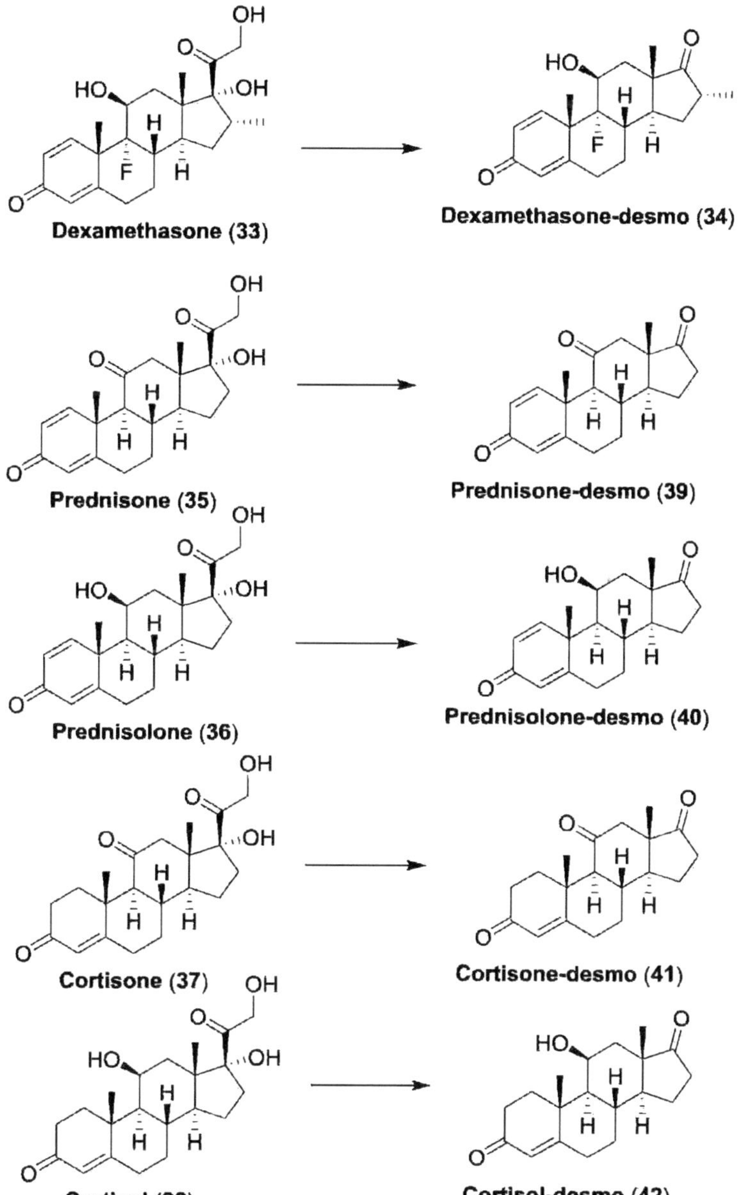

Figure 9.
Gut bacterial metabolism of corticosteroid drugs, dexamethasone (33), prednisone (35), prednisolone (36), cortisone (37), and cortisol (38), to their respective products ***34****,* ***39****,* ***40****,* ***41****, and* ***42*** *via the desmolytic activity.*

Systematic identification of drug-metabolizing genes encoded by gut bacteria provides the mechanistic insights into drug metabolism in human [62]. Genes of the gut bacterium *Bacteroides thetaiotaomicron* were cloned into *Escherichia coli*, leading to the identification of new 16 gene products, which were able to metabolite 18 drugs to 41 different metabolites [62]. Certain gene products have specificity and cross-activity, and gene deletion and complementation techniques revealed the mechanisms of individual gene products. For instance, the *bt2068* gene encodes the enzyme that could reduce (adding H_2) norethisterone acetate (**24**) (**Figure 8**), as well as other similar steroid drugs such as levonorgestrel and progesterone, while the *bt2367* gene encodes acyltransferase that converts the drug pericyazine (**43**) to both acetyl- and propionyl-pericyazine products, e.g., acetyl-*O*-pericyazine (**44**) and propionyl-*O*-pericyazine (**45**), respectively (**Figure 10**) [62]. It is known that the metabolism products of a drug diltiazem (**46**) are *N*-desmethyldiltiazem (**47**), *N*,*N*-didesmethyldiltiazem (**48**), *O*-desmethyldiltiazem (**49**), *N*,*O*-didesmethyldiltiazem (**50**), desacetyldiltiazem (**51**), desacetyl-*N*-desmethyldiltiazem (**52**), desacetyl-*N*,*N*-didesmethyldiltiazem (**53**), desacetyl-*O*-desmethyldiltiazem (**54**), and desacetyl-*N*,*O*-didesmethyldiltiazem (**55**) (**Figure 10**) [72]. The gene *bt4096* in gut bacteria is responsible for the deacetylation of diltiazem (**46**) and its metabolites **47–50** to give their corresponding deacetylated products **51–55**, respectively (**Figure 10**) [62]. This study suggests that gut bacteria substantially contribute to drug metabolism in the human gastrointestinal tract, and the metabolism of drug candidates by gut microbiome should be studied as a part of the drug development processes.

The drug metabolism by gut microbiome can give negative effects to drug efficacy, thus leading to the decrease in efficiency and potency of certain drugs. L-dopa or levodopa (**56**) (**Figure 11**) is the first-line drug for the treatment of Parkinson's disease; the metabolism of this drug by gut microbiome provides negative effects for Parkinson's patients. The drug L-dopa (**56**) can cross the blood–brain barrier, entering the central nervous system and then transforming to a neurotransmitter,

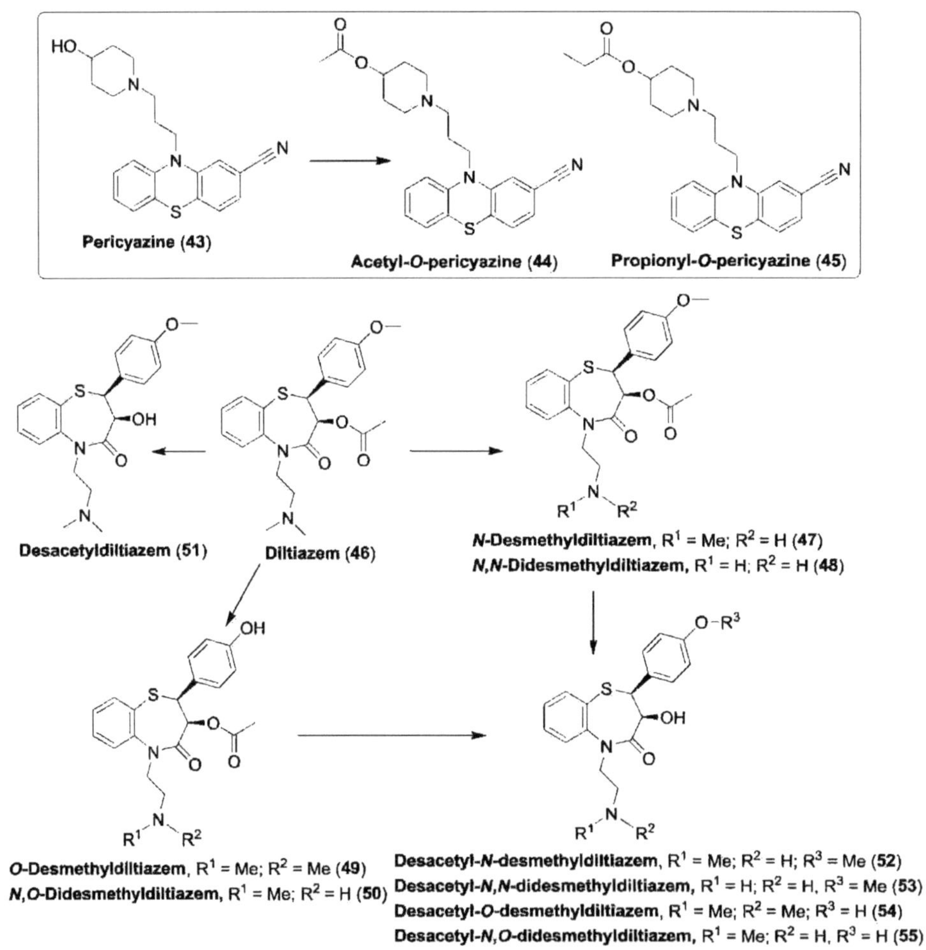

Figure 10.
Metabolism of pericyazine (43) to acetyl-O-pericyazine (44) and propionyl-O-pericyazine (45) and metabolism of diltiazem (46) to the metabolite products 47–55 by gut bacteria.

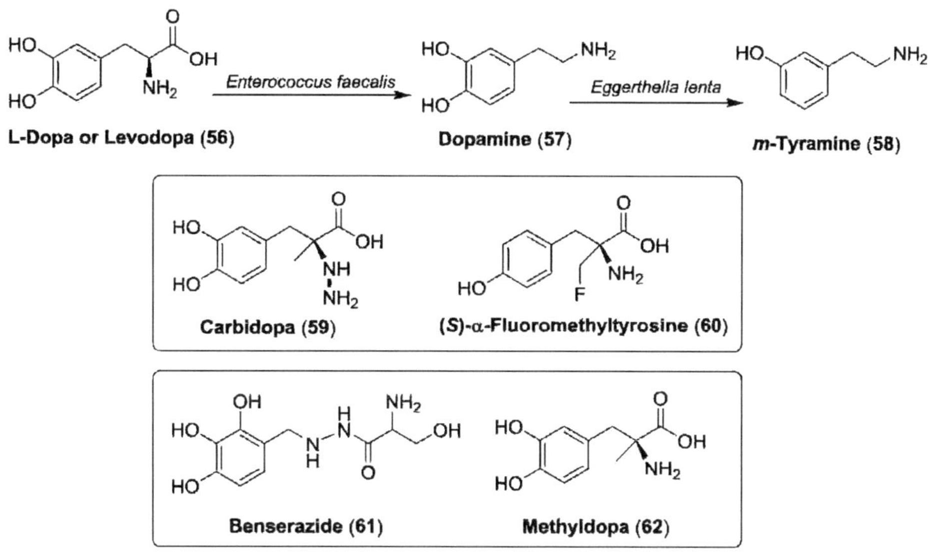

Figure 11.
Bioconversion of L-dopa (56) to dopamine (57) and m-*tyramine (58) by gut bacteria and structures of inhibitors of amino acid decarboxylases, carbidopa (59), (S)-α-fluoromethyltyrosine (60), benserazide (61), and methyldopa (62).*

dopamine (**57**), by the enzyme pyridoxal phosphate L-amino acid decarboxylase (**Figure 11**). It is known that intestinal microflora (gut microbiome) can metabolite L-dopa (**56**) [73] and that the formation of *m*-tyramine (**58**) from L-dopa (**56**) is a side effect of this drug for parkinsonism (**Figure 11**) [74]. A neurotransmitter, dopamine (**57**), is the only active agent needed for the treatment of parkinsonism, and it should be formed from L-dopa (**56**) at the central nervous system after L-dopa (**56**) crossing the blood–brain barrier. However, the generation of dopamine (**57**) from the drug L-dopa (**56**) can occur at the human gastrointestinal tract (known as peripheral metabolism), not at the central nervous system, thus giving undesirable side effects. To prevent this peripheral metabolism, an inhibitor of pyridoxal phosphate L-amino acid decarboxylase, carbidopa (**59**) (**Figure 11**), is coadministered with the drug L-dopa (**56**). It is known for many years that microorganisms can decarboxylate L-dopa (**56**) to dopamine (**57**), which in turn undergoes the dehydroxylation reaction to give *m*-tyramine (**58**) [75]. The treatment of L-dopa (**56**) is improved when patients receive broad-spectrum antibiotics, which suppress the growth of gut bacteria, indicating that gut bacteria are involved in the decrease of therapeutic efficiency of L-dopa (**56**) [76]. Therefore, gut microbiome can potentially reduce the drug efficacy of L-dopa (**56**) through their metabolic activities toward the drug.

Recent study led by Prof. Balskus revealed that the gut bacterium *Enterococcus faecalis* has the *tyrDC* gene encoding the enzyme tyrosine decarboxylase that is able to decarboxylate both L-dopa (**56**) and an amino acid, tyrosine [77]. Moreover, the gut bacterium *Eggerthella lenta* has the *dadh* gene encoding a molybdenum cofactor-dependent dopamine dehydroxylase, which is the enzyme responsible for the dehydroxylation of dopamine (**57**) to *m*-tyramine (**58**) (**Figure 11**). The metabolism of L-dopa (**56**) and dopamine (**57**) in complex gut microbiotas of Parkinson's patients is dependent on the *tyrDC* and *dadh* genes [77]. The study demonstrated that carbidopa (**59**), an inhibitor of pyridoxal phosphate L-amino acid decarboxylase, failed to prevent L-dopa (**56**) metabolism in complex gut microbiotas of Parkinson's patients [77]. However, another inhibitor, (*S*)-α-fluoromethyltyrosine

(**60**) (**Figure 11**), could prevent the decarboxylation of L-dopa (**56**) that is from the metabolic activities of both the bacterium *E. faecalis* and complex gut microbiotas of Parkinson's patients [77]. In a mouse model, levels of the drug L-dopa (**56**) increased in serum when coadministered (*S*)-α-fluoromethyltyrosine (**60**) with L-dopa (**56**) to mice colonized with the gut bacterium *E. faecalis* [77]. An independent study led by Prof. Aidy also identified the *tdc* gene responsible for tyrosine decarboxylases in the gut bacterium *E. faecium*; bacterial tyrosine decarboxylases efficiently convert the drug L-dopa (**56**) to dopamine (**57**) [78]. Aidy and co-workers also found that carbidopa (**59**) in L-dopa (**56**) combination therapy did not inhibit the activities of decarboxylase enzymes in gut bacteria, *E. faecalis* and *E. faecium* [78]. Moreover, other decarboxylase inhibitors, benserazide (**61**) and methyldopa (**62**) (**Figure 11**), also failed to inhibit the decarboxylase activity of gut bacteria toward the drug L-dopa (**56**). These studies demonstrated gut microbiota significantly reduced the levels of the drug L-dopa (**56**) in a body, thus contributing to the higher dosages required for the Parkinson's patients that have gut microbiome with high metabolism toward the drug L-dopa (**56**). Variations in gut microbiota among Parkinson's patients might contribute to the different responses, i.e., harmful side effects and decreased efficacy, to the drug L-dopa (**56**). Therefore, gut microbiota plays a critical role in the drug metabolism and considerably contributes to treatment outcomes of this drug.

Gut microbiota can improve therapeutic efficiency of certain drugs for particular treatments. Cancer immunotherapy is relatively new for cancer treatment using human immune system to control and eradicate cancer cells, and it is more precise and personalized, thus providing more effectiveness with fewer side effects than other cancer treatments. Gut microbiota was found to play a role in cancer immunotherapy targeting CTLA-4, a protein receptor downregulating the immune system, because anticancer effects of CTLA-4 blockade were found to depend on gut bacteria of *Bacteroides* species, e.g., *Bacteroides thetaiotaomicron* or *B. fragilis* [79]. The study demonstrated that germ-free mice did not show the response to CTLA blockade, thus defecting an anticancer property of the drug. Indeed, this drug deficiency could be improved by gavage with the gut bacterium *B. fragilis* through immunization with the bacterium polysaccharides or by adoptive transfer of *B. fragilis*-specific T cells. Therefore, this research study demonstrates that the gut bacterium could help patients treated with a monoclonal antibody drug for the treatment of cancer targeting CTLA-4 [79].

Gut microbiome also improves therapeutic effect of a cancer immunotherapy targeting immune checkpoint inhibitor via the PD-1/PD-L1 pathway [59]. Antibiotics are found to give negative effects for patients treated with cancer immunotherapies as they inhibit the efficacy of immune checkpoint inhibitor drug that targets the programmed cell death receptor of the PD-1/PD-L1 pathway [59]. Moreover, suppression of growth of gut bacteria by antibiotic drugs leads to the decrease of drug efficacy, suggesting that gut microorganisms are important for this cancer therapy. The study demonstrated that gut microbiota provided significant effects on cancer immunotherapies targeting the PD-1/PD-L1 interaction because there was substantial association between commensal microorganisms and therapeutic response of anticancer drug that inhibits the activity of PD-1 and PD-L1 immune checkpoint proteins [80]. Gut bacteria including *Collinsella aerofaciens*, *Enterococcus faecium*, and *Bifidobacterium longum* were found to be associated with the improvement of this cancer immunotherapy. Intriguingly, reconstitution of germ-free mice with fecal samples from patients with good drug response could help to control tumor, leading to better efficacy of anti-PD-L1 cancer therapy [80]. An independent study revealed that the gut bacterium *Akkermansia muciniphila* assists cancer immunotherapy targeting the PD-1/PD-L1 interaction toward epithelial tumors [81]. The study on fecal microbiota transplantation demonstrated that germ-free or antibiotic-treated mice receiving gut bacteria from patients with good response to cancer immunotherapy have significant therapeutic improvement, while those receiving the samples from nonresponding patients do not have such improvement for cancer immunotherapy [81]. Restoration of the drug efficacy in germ-free mice receiving the samples from nonresponding patients was simply achieved by oral supplementation with the gut bacterium *A. muciniphila*, indicating the benefit of gut microbiota for this cancer immunotherapy. Another independent research also found similar benefits of gut microbiota on anti-PD-1 immunotherapy in melanoma patients; this study investigated microbiome samples from 112 patients with metastatic melanoma and found that there were substantial differences in the composition and diversity of gut microbiome obtained from patients with good drug response and from nonresponding patients [82]. Patients with good response to immunotherapeutic PD-1 blockade have abundance of gut bacteria of the family *Ruminococcaceae* and *Faecalibacterium*, while patients with poor response to immunotherapeutic PD-1 blockade tend to have relative abundance of *Bacteroidales*. It is possible that patients with a favorable gut microbiome, e.g., *Ruminococcaceae* and *Faecalibacterium*, toward the immunotherapeutic PD-1 blockade therapy have improved systemic and immune responses mediated by certain factors such as improvement of effector T cell function in the periphery, increase of antigen production, and improvement of the tumor microenvironment [82].

Gut microbiota also has an important role in chemotherapy for cancer treatment because they can modulate drug efficacy, for example, eliminating the anticancer properties of the drug or mediating toxicity [83]. Cyclophosphamide (**63**) (**Figure 12**) is a drug used in cancer chemotherapy for many types of cancers, as well as for autoimmune diseases, and its mechanism is through the stimulation of anticancer immune responses. In a mouse model, the composition of gut microbiota is changed after administration of cyclophosphamide (**63**), and this drug induces the translocation of certain Gram-positive bacteria into secondary lymphoid organs. Gut bacteria could stimulate certain immune responses beneficial to cancer therapy. Germ-free mice carrying tumor treated with antibiotics to kill Gram-positive bacteria had less therapeutic response, and their tumors exhibited resistance to the drug cyclophosphamide (**63**), suggesting that gut microbiota improves anticancer immune response [84]. Gut bacteria, *Enterococcus hirae* and *Barnesiella intestinihominis*, were found to help cyclophosphamide-induced therapeutic immunomodulatory response, thus improving the efficacy of this alkylating immunomodulatory drug [85].

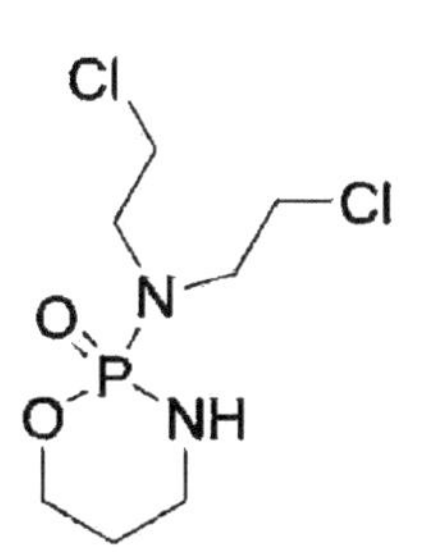

Cyclophosphamide (63)

Figure 12.
Structure of an anticancer drug, cyclophosphamide (63).

The research studies mentioned earlier demonstrate the interactions of drugs and gut microbiome that provide beneficial effects on cancer therapy. However, interactions of gut microbiome and drugs can also give negative influence in cancer treatment, for example, the treatment of an anticancer drug, gemcitabine (**64**) or 2′,2′-difluorodeoxycytidine (**Figure 13**), which is a derivative of cytidine nucleoside base. A research led by Straussman showed that the bacterium *Mycoplasma hyorhinis* was found to be the cause of gemcitabine (**64**) resistance in colon carcinoma models [86]. In a colon cancer mouse model, *M. hyorhinis* could metabolize gemcitabine (**64**) to the corresponding deaminated derivative, 2′,2′-difluorodeoxyuridine (**65**), that does not have anticancer activity (**Figure 13**). The conversion of gemcitabine (**64**) to 2′,2′-difluorodeoxyuridine (**65**) was previously reported [87], and the nucleoside-catabolizing enzymes, i.e., cytidine deaminase, in the bacterium, *M. hyorhinis*, were also identified [88]. Straussman and co-workers analyzed genes and genomes of 2674 bacterial species and found that most of the *Gammaproteobacteria* class had the gene coding for the enzyme cytidine deaminase, thus potentially mediating gemcitabine resistance [86]. In a mouse model of colon carcinoma, mice receiving an antibiotic, ciprofloxacin, showed a good response to the anticancer drug gemcitabine (**64**), indicating that suppression of the growth of certain bacteria led to the improvement of the drug efficacy. Investigation of human pancreatic ductal adenocarcinoma collected from pancreatic cancer surgery revealed that there were intratumor bacteria, mainly belonging to the class *Gammaproteobacteria* such as *Enterobacteriaceae* and *Pseudomonadaceae* families in these samples; the intratumor bacteria can mediate resistance to chemotherapy of the drug gemcitabine (**64**) [86]. Therefore, the metabolism of the drug gemcitabine (**64**) to 2′,2′-difluorodeoxyuridine (**65**) by gut microbiota provides the negative effects for cancer treatment. This study underscores the importance of the research on drug metabolism by gut microbiome, which should be investigated for the new drug candidates during the drug development processes.

Additional example for the negative effects of gut microbiota for cancer chemotherapy is the treatment of colorectal cancer with the drugs, oxaliplatin (**66**) and fluorouracil or 5-FU (**67**) (**Figure 13**); the gut bacterium *Fusobacterium*

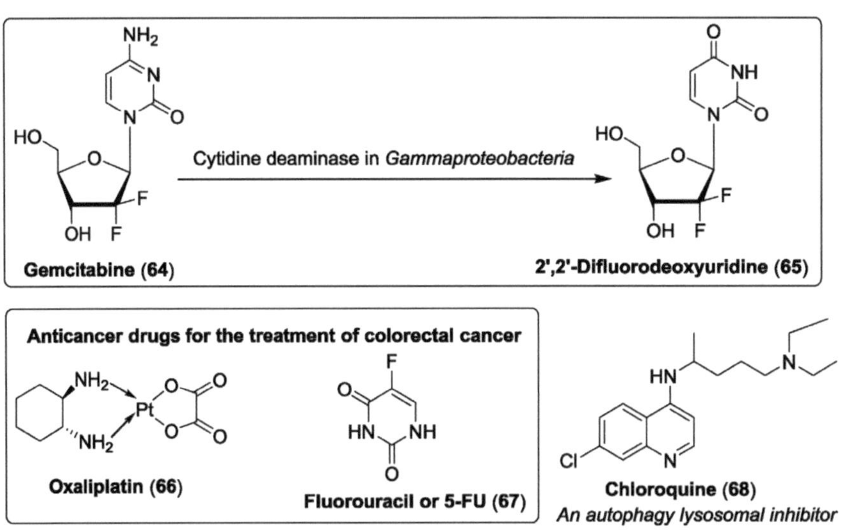

Figure 13.
Biotransformation of gemcitabine (64) to its metabolite 2′,2′-difluorodeoxyuridine (65); structures of anticancer drugs, oxaliplatin (66) and fluorouracil or 5-FU (67), and autophagy lysosomal inhibitor, chloroquine (68).

nucleatum was found to promote resistance to chemotherapy for colorectal cancer [89]. Analysis of colorectal cancer tissues collected from patients with recurrence or without recurrence of cancer revealed that the bacterium *F. nucleatum* is associated with the recurrence of colorectal cancer, which is derived from chemoresistance toward the drugs [89]. Cultivation of colorectal cancer cells co-cultured with *F. nucleatum* revealed that the bacterium potentially activated an autophagy pathway in colorectal cancer cells. An addition of a known autophagy lysosomal inhibitor, chloroquine (**68**) (**Figure 13**), could inhibit autophagic flux in the *F. nucleatum*-cultured cells, confirming the autophagy activation induced by the gut bacterium *F. nucleatum* [89]. Moreover, this bacterium reduced cell apoptosis of colorectal cancer cells, indicating that it specifically induced resistance toward the drugs oxaliplatin (**66**) and fluorouracil (**67**). Co-cultured cancer cells with the bacterium *F. nucleatum* and treated cancer cells with the drugs oxaliplatin (**66**) and fluorouracil (**67**) in the presence of autophagy lysosomal inhibitor, chloroquine (**68**), could eradicate chemoresistant effect, strongly confirming that the bacterium *F. nucleatum* induced chemoresistance through the autophagy pathway [89]. Detailed mechanistic study revealed that the bacterium *F. nucleatum* mediated chemoresistance through the TLR4 and MYD88 signaling pathway [89]. An independent study showed that the gut bacterium *F. nucleatum* is a diagnostic marker of colorectal cancer because patients with this cancer generally have high density of this bacterium in tumor cells [90]. Several studies have shown the prevalence of the bacterium *F. nucleatum* in colorectal tissues and fecal samples of patients, and those with high density of this bacterium tend to have lower rate of survival [91]. Therefore, manipulation of the bacterial population of *F. nucleatum* might be useful for the treatment of colorectal cancer, and this bacterium is potentially a diagnostic and/or prognostic marker for colorectal cancer.

In addition to drug metabolism, gut microbiota is also involved in drug–drug interactions when patients take two drugs at the same time, particularly when using antibiotics together with other drugs. Several studies have demonstrated the effects of antibiotic drugs on the metabolic activities of gut microbiota toward drugs and phytochemicals [92]. An example of a drug–drug interaction is the contribution of an antibiotic drug, amoxicillin (**69**), to a nonsteroidal anti-inflammatory drug aspirin (**70**) (**Figure 14**) [93]. It is worth mentioning that aspirin (**70**) is used not only for a pain reliever but also for primary prevention of cardiovascular disease [94] and cancer chemoprevention [95]. Recent study showed that amoxicillin (**69**) potentially affected the composition of gut microbiota by reducing number and species of intestinal bacteria in rats; the abundance of the gut bacteria, *Prevotella copri* and *Helicobacter pylori*, was reduced significantly after rats receiving amoxicillin (**69**) [93]. Gut microorganisms in rats could metabolite aspirin (**70**) to salicylate or salicylic acid (**71**) (**Figure 14**). Salicylate is a conjugate base of salicylic acid (**71**). It is known that the drug aspirin (**70**) is not responsible for a pain relief, but its metabolite, salicylic acid (**71**), is the active metabolite responsible for a pain relief with anti-inflammatory effect [96]. Therefore, gut microbiota plays an important role in the biotransformation of the drug aspirin (**70**) into the active metabolite, salicylic acid (**71**). After an oral administration of an antibiotic drug amoxicillin (**69**) to rats, the reduction of the metabolism of aspirin (**70**) into salicylic acid (**71**) was observed, suggesting the decrease of gut microbiota by amoxicillin (**69**) led to the reduction of the biotransformation of aspirin (**70**) into salicylic acid (**71**). Further study on the pharmacokinetics of aspirin (**70**) in rats revealed that amoxicillin (**69**) significantly affected the pharmacokinetic properties of aspirin (**70**) [93]. This study indicates that changes of the composition of gut microbiome by antibiotic drugs could substantially disturb the therapeutic effect of other drugs.

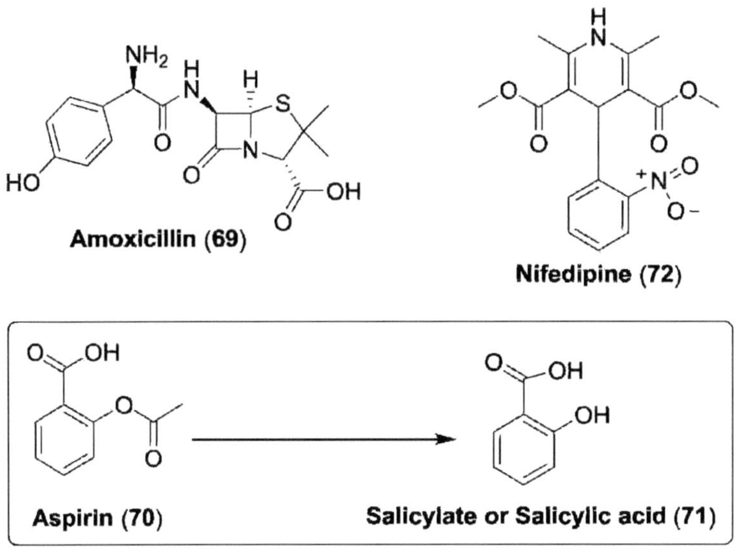

Figure 14.
Structures of amoxicillin (69), aspirin (70) and its metabolite, salicylate or salicylic acid (71), and nifedipine (72).

Previous study also showed that antibiotics substantially reduced the metabolic activity of gut microbiota toward aspirin (**70**), leading to the reduction of an antithrombotic effect of aspirin (**70**) [97]. Moreover, environmental changes, e.g., high-altitude hypoxia, also give effects on the pharmacokinetics and pharmacodynamics of aspirin (**70**) because of the changes in gut microbiota [98]. In an animal model, the plateau hypoxic environment affected the composition of gut microbiome because it increased the bacterial species of *Bacteroides* in rat feces but reduced numbers of the bacteria of the genus *Prevotella*, *Coprococcus*, and *Corynebacterium*. Changes in gut microbiome affected the metabolism of aspirin (**70**), thus altering the bioavailability of aspirin (**70**) in patients [98]. Plateau hypoxic environment also has the effects on the drug nifedipine (**72**), which could be metabolized by gut microorganisms (**Figure 14**) [99]. Nifedipine (**72**) is a drug for the treatment of hypertension, precordial angina, and certain vascular diseases. Plateau hypoxic environment was found to alter the composition of gut microbiota in an animal model, thus affecting the bioavailability of nifedipine (**72**) [99].

Recent study led by Kittakoop revealed that valproic acid or valproate (**73**) (**Figure 15**), an anticonvulsive drug used for treatments of epilepsy and bipolar disorder, had effects on the biosynthesis of fatty acids in microorganisms including representative gut microbiome [100]. Valproic acid (**73**) is also an epigenetic modulator, acting as an inhibitor of histone deacetylase [101]. Initially, Kittakoop and co-workers employed "One strain many compound" (OSMAC) approach using the marine fungus *Trichoderma reesei* treated with an epigenetic modulator, valproic acid (**73**), aiming to modulate the fungus *T. reesei* to produce new natural products, which are secondary metabolites. However, valproic acid (**73**) was found to have the effects on the biosynthesis of fatty acids, which are primary metabolites, instead of natural products that are secondary metabolites [100]. The study revealed that valproic acid (**73**) at a concentration of 100 μM could either inhibit or induce the biosynthesis of certain fatty acids in fungi, yeast, and bacteria. Valproic acid (**73**) inhibited the biosynthesis of palmitoleic acid (C16:1), α-linolenic acid (C18:3), arachidic acid (C20:0), and lignoceric acid (C24:0) in the fungus *Fusarium oxysporum*, while it induced the production of α-linolenic acid (C18:3) in the fungus *Aspergillus*

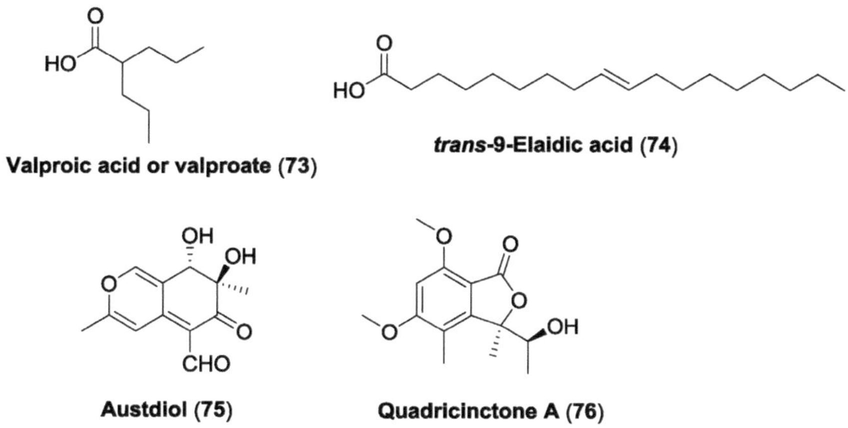

Figure 15.
Structures of valproic acid or valproate (73), trans*-9-elaidic acid (74), austdiol (75), and quadricinctone A (76).*

aculeatus [100]. The bacterium of the genus *Pediococcus* is commonly found as gut microbiome in humans and animals [102]; valproic acid (**73**) was found to inhibit the production of lignoceric acid (C24:0) in the bacterium, *Pediococcus acidilactici* [100]. The yeast *Candida utilis* was found as gut microbiome in pediatric patients with inflammatory bowel disease [103]; valproic acid (**73**) inhibited the biosynthesis of palmitoleic acid (C16:1) and α-linolenic acid (C18:3) in *C. utilis* [100]. The yeast *Saccharomyces cerevisiae* was previously found as a prevalent gut microbiome in human [104], and the drug valproic acid (**73**) was found to inhibit the production of *α*-linolenic acid (C18:3) in the yeast *S. cerevisiae* [100]. Interestingly, valproic acid (**73**) could induce the biosynthesis of *trans*-9-elaidic acid (**74**) (**Figure 15**) in the yeast *Saccharomyces ludwigii* [100]. In human, *trans*-9-elaidic acid (**74**) could increase intracellular Zn^{2+} in macrophages and inhibit *β*-oxidation in peripheral blood macrophages [105, 106]; this suggests that the production of *trans*-9-elaidic acid (**74**) in gut microorganisms induced by the drug valproic acid (**73**) may indirectly give the effects to human. Valproic acid (**73**) also had effects on the biosynthesis of polyketides because it substantially reduced the production of austdiol (**75**) (90% reduction) and quadricinctone A (**76**) (50% reduction) (**Figure 15**), which are the polyketides of the fungus *Dothideomycetes* sp. [100]. The biosynthesis of fatty acids is considerably similar to that of polyketides, i.e., sharing the same catalytic roles and biosynthetic precursors [107]. Therefore, the drug valproic acid (**73**) possibly gives effects on the biosynthetic pathways of both fatty acids and polyketides because of their biosynthetic similarities. Gut microbes have biosynthetic gene clusters involving in the biosynthesis of many bioactive natural products including polyketides [108]; some natural products produced by gut microbiome have biological activities. This study suggests that commonly used drugs could potentially give the effects on the biosynthesis of secondary metabolites (natural products) of gut microbiome.

3.2 Interactions of gut microbiome and natural products

Traditional medicine and natural products have significant interactions with gut microbiome. Many studies revealed that dietary natural products modulating gut microbiota are useful for prevention and management of diabetes mellitus [109]. Recent study revealed that a traditional Chinese herbal formula and an antidiabetic drug, metformin (77) (**Figure 16**), could improve the treatment of

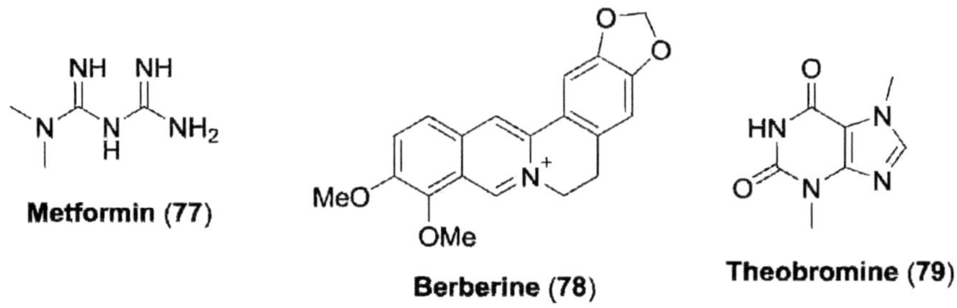

Figure 16.
Structures of metformin (77), berberine (78), and theobromine (79).

type 2 diabetes with hyperlipidemia by enriching certain beneficial species of gut bacteria, for example, *Faecalibacterium* sp. and *Blautia* [110]. The study was carried out in 450 patients with type 2 diabetes and hyperlipidemia, and the profiles of gut microbiota were analyzed using fecal samples in patients treated with metformin and a traditional Chinese herbal formula. An antidiabetic drug metformin (**77**) and herbal medicine significantly changed the gut microbiota profile that led to the enhancement of therapeutic effects of the drugs [110]. The traditional Chinese herbal formula used in the study contains the plants including *Coptis chinensis*, *Momordica charantia*, *Rhizoma anemarrhenae*, and *Aloe vera*, as well as red yeast rice from the fermentation; this herbal recipe is practically used in clinical application [110]. Among the plants used in this formula, *Coptis chinensis* contains an alkaloid berberine (**78**) (**Figure 16**). An independent study revealed that berberine (**78**) could significantly change the composition of gut microbiota in high-fat diet-fed rats [111]. An alkaloid berberine (**78**) was able to enrich selectively short-chain fatty acid-producing bacteria such as the genus *Blautia* and *Allobaculum* [111]. Another independent study also revealed that both metformin (**77**) and berberine (**78**) could change profiles of gut microbiota in high-fat diet-induced obesity in rats [112]. Substantial reduction of the diversity of gut microbiota was observed by both metformin (**77**) and berberine (**78**) because 60 out of the 134 operational taxonomic units were decreased after treatment with both drugs. However, there were considerable increases in short-chain fatty acid-producing bacteria, e.g., the genus *Butyricicoccus*, *Blautia*, *Allobaculum*, *Phascolarctobacterium*, and *Bacteroides*, after treatment with both metformin (**77**) and berberine (**78**) [112]. Therefore, in addition to the direct benefit toward the treatment of diabetes and obesity, the drugs, metformin (**77**) and berberine (**78**), could also improve gut microbiota profile by increasing short-chain fatty acid-producing bacteria and thus mediating their useful effects on the host [112]. As mentioned earlier in Section 2, gut microbiomes that produce short-chain fatty acids provide many beneficial effects on human health [24, 25].

Recent study revealed that berberine (**78**) could prevent ulcerative colitis by modifying gut microbiota and regulating T regulatory cell and T helper 17 cell in a dextran sulfate sodium-induced ulcerative colitis mouse model [113]. The diversity of gut microbiota was reduced by berberine (**78**), which markedly interfered the abundance of certain bacterial genus such as *Bacteroides*, *Desulfovibrio*, and *Eubacterium*. Therefore, the mechanisms of berberine (**78**) for the prevention of ulcerative colitis are by regulating the balance of T regulatory cell and T helper 17 cell, as well as by modifying gut microbiota [113]. Theobromine (**79**) (**Figure 16**) is a xanthine alkaloid of cocoa beans and found in chocolate, and its structure is closely related to caffeine. A cocoa-enriched diet containing theobromine (**79**) could decrease the intestinal immunoglobulin A secretion and immunoglobulin

A-coating bacteria, i.e., the genus of *Bacteroides*, *Staphylococcus*, and *Clostridium* [114]. A cocoa-enriched diet had effects on a differential toll-like receptor pattern, which led to changes in the intestinal immune system [114]. Moreover, further experiments in rats revealed that a diet containing 10% cocoa and a diet supplemented with 0.25% theobromine (**79**) could reduce the gut bacterium *Escherichia coli*, while a diet with 0.25% theobromine (**79**) reduced the gut bacterial community of *Clostridium histolyticum*, *C. perfringens*, *Streptococcus* sp., and *Bifidobacterium* sp. [115]. The amounts of short-chain fatty acids increased after feeding rats with a diet containing 10% cocoa and that supplemented with 0.25% theobromine (**79**), while both diets decreased the abundance of immunoglobulin A (IgA)-coated bacteria. It is worth mentioning that gut IgA-coated bacteria could potentially cause intestinal disease such as inflammatory bowel disease, and eradication of these bacteria may prevent or reduce intestinal disease development [116]. Therefore, the active natural product theobromine (**79**) in cocoa able to reduce the amounts of immunoglobulin A-coated bacteria, and to modify the profile of gut microbiota, provides beneficial effects on human health [115].

It is known that berberine (**78**) has poor solubility; however, it can show effectiveness for the treatment of certain diseases; therefore, there might be a specific mechanism to deliver berberine (**78**) to an organ system. In an animal model, berberine (**78**) was found to convert to dihydroberberine (**80**) in an intestinal ecosystem of rats (**Figure 17**); the metabolite dihydroberberine (**80**) exhibited much better absorption rate than its parent drug, berberine (**78**) [117]. Incubation of berberine (**78**) with human gut bacteria isolated from gastrointestinal human specimens also produced dihydroberberine (**80**), and the amounts of dihydroberberine (**80**) obtained from the biotransformation of gut bacteria were higher than that obtained from other bacteria, which were not gut bacteria and used as the control. This experiment confirmed that gut microbiota could convert berberine (**78**) into its absorbable form, dihydroberberine (**80**); therefore intestinal microbiota is considered to be a "tissue" or an "organ" that is able to transform berberine (**78**)

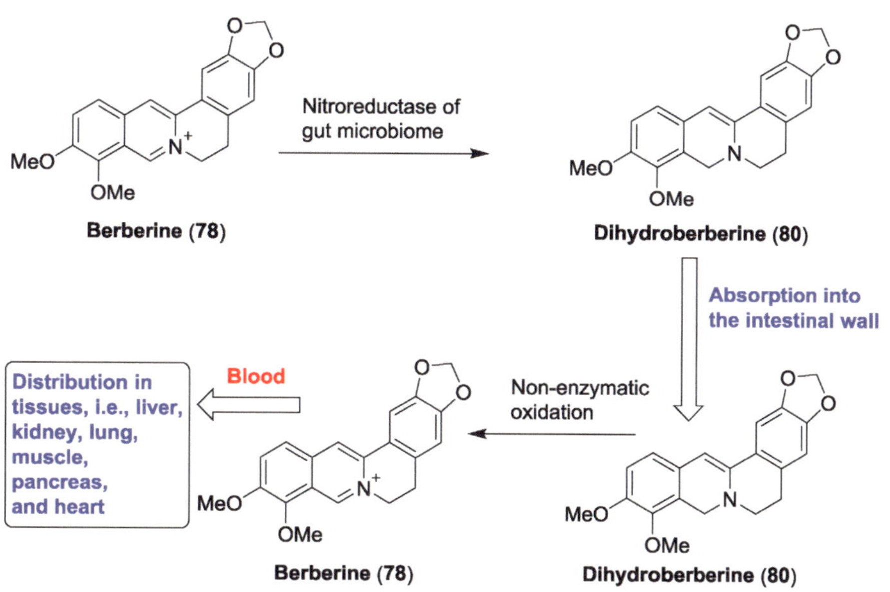

Figure 17.
Bioconversion of berberine (78) into an absorbable form, dihydroberberine (80), by gut bacteria; absorption of dihydroberberine (80) into the intestine wall and nonenzymatic conversion of dihydroberberine (80) to the active form berberine (78).

into an absorbable form, dihydroberberine (**80**) [117]. Mechanistic study revealed that gut microbiome uses the enzyme nitroreductases to catalyze the conversion of berberine (**78**) to dihydroberberine (**80**) (**Figure 17**). Dihydroberberine (**80**) was found to be absorbed in intestinal epithelia, but it was reverted to the active form berberine (**78**) soon after entering tissues of the intestinal wall. Detailed analysis showed that the conversion of dihydroberberine (**80**) back to berberine (**78**) was by a nonenzymatic oxidation through multi-faceted factors, for example, superoxide anion and metal ions, which occurred in intestinal epithelial tissues (**Figure 17**) [117]. Previous report demonstrated that dihydroberberine (**80**) in its sulfate form, e.g., dihydroberberine sulfate, also showed better absorption in the intestine than its parent drug, berberine (**78**) [118]. Recent independent studies revealed that dihydroberberine (**80**) has interesting biological activities, for example, anti-inflammatory activity through dual modulation of NF-κB and MAPK signaling pathways [119], synergistic effects with an anticancer drug sunitinib on human non-small cell lung cancer cell lines by inflammatory mediators and repressing MAP kinase pathways [120], and inhibition of ether-a-go-go-related gene (hERG) channels expressed in human embryonic kidney 293 (HEK293) cells [121]. It is worth mentioning that the metabolite products from gut biomicrobiota, i.e., dihydroberberine (**80**), have different biological activity from its parent drug, berberine (**78**). Therefore, the drug development process should include a research study on the metabolism of natural products (as drug candidates) by gut microbiota.

Demethyleneberberine (**81**), berberrubine (**82**), jatrorrhizine (**83**), and thalifendine (**84**) were found as major metabolites in rats after an oral administration of berberine (**78**) (**Figure 18**) [122]. Comparison of the levels of these metabolites in conventional rats (a control group) and pseudo germ-free rats revealed that liver and intestinal bacteria were involved in the metabolism and disposition of berberine (**78**) in vivo. It is worth mentioning that some metabolites from this biotransformation exert important biological activities. For example, demethyleneberberine (**81**) inhibits oxidative stress, steatosis, and mitochondrial dysfunction in a mouse

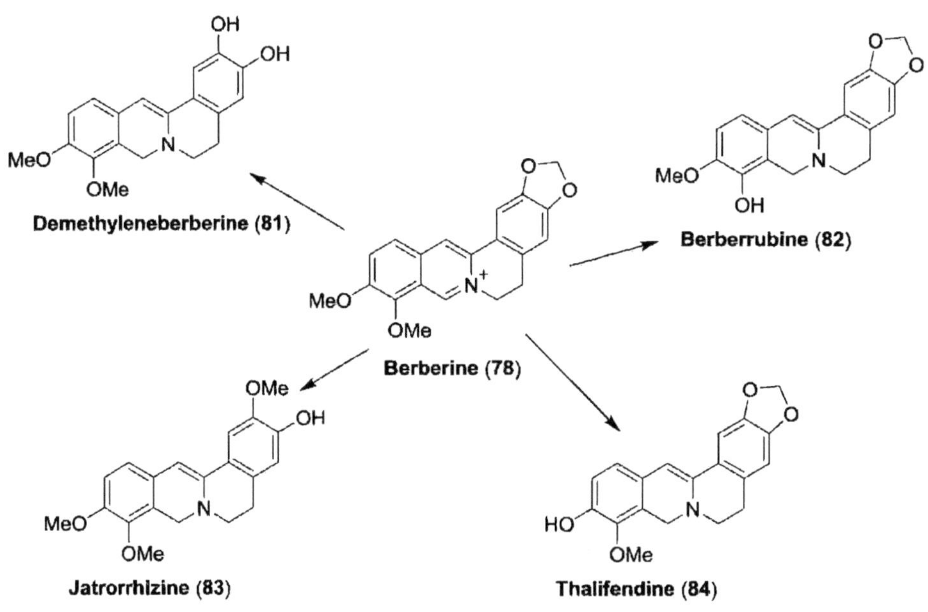

Figure 18.
Biotransformation of berberine (78) to demethyleneberberine (81), berberrubine (82), jatrorrhizine (83), and thalifendine (84) by gut bacteria.

model, which is a potential therapy for alcoholic liver disease [123]. Berberrubine (**82**) was found to reduce inflammation and mucosal lesions in dextran sodium sulfate-induced colitis in mice, which might be useful for the treatment of ulcerative colitis [124]. Jatrorrhizine (**83**) could reduce the uptake of 5-hydroxy-tryptamine and norepinephrine by the inhibition of uptake-2 transporters, thus exerting antidepressant-like action in mice [125]. Therefore, the biotransformation of berberine (**78**) by gut bacteria leads to the production of bioactive metabolites, which have interesting pharmacological properties; this underscores the impact of gut microbiota in the drug development process for natural products.

Since there are interactions between gut microbiota and natural products, efforts have been made to use natural compounds for the treatment of gut microbiota dysbiosis, which is the imbalance of microorganisms in the human gastrointestinal tract. Dysbiosis of gut microbiota is strongly associated with some diseases such as type 2 diabetes, inflammatory bowel disease, obesity, and nonalcoholic fatty liver disease [16, 126]. Alkaloids of a medicinal plant, *Corydalis saxicola*, were used to prevent gut microbiota dysbiosis in an animal model [127]. Major alkaloids in *Corydalis saxicola* are berberine (**78**), jatrorrhizine (**83**), dehydrocavidine (**85**), palmatine (**86**), and chelerythrine (**87**) (**Figures 18** and **19**). Among these alkaloids, berberine (**78**), palmatine (**86**), and chelerythrine (**87**) are the main active principles for the treatment of antibiotic-induced gut microbiota dysbiosis through the key enzyme, CYP27A1, which is involved in the biosynthesis of bile acid [127]. This study provides insights for the discovery of natural products for the treatment of gut microbiota dysbiosis.

Xanthohumol (**88**) is a prenylflavonoid in hops (*Humulus lupulus*), which is responsible bitter flavor in beer (**Figure 20**). Xanthohumol (**88**) has interesting pharmacological properties, for example, improving cognitive flexibility in young mice [128] and having beneficial effects toward metabolic syndrome-related diseases such as type 2 diabetes and obesity [129]. The comparative study on germ-free and human microbiota-associated rats toward the metabolism of xanthohumol (**88**) revealed that gut bacteria could transform xanthohumol (**88**) to isoxanthohumol (**89**) and 8-prenylnaringenin (**90**), respectively (**Figure 20**) [130]. The metabolism of xanthohumol (**88**) was further studied using human intestinal bacteria, *Eubacterium ramulus* and *E. limosum*. It is worth mentioning that an independent study revealed that the bacteria of the genus *Eubacterium* are normally abundant in the human gastrointestinal tract; their densities in human gut are up to 10^{10} colony-forming units/g of intestinal content [131]. Xanthohumol (**88**) is spontaneously converted to isoxanthohumol (**89**), which is in turn bioconverted to 8-prenylnaringenin (**90**) by the gut bacterium, *E. limosum* (**Figure 20**) [132]. 8-Prenylnaringenin (**90**) is biotransformed to *O*-desmethylxanthohumol (**91**) by the bacterium *E. ramulus*; this bacterium could also convert *O*-desmethylxanthohumol (**91**) to desmethyl-α,β-dihydroxanthohumol (**92**). Moreover, the bacterium *E. ramulus* was

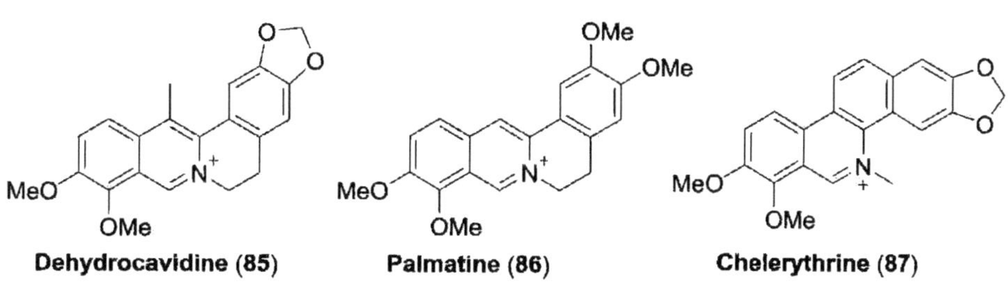

Figure 19.
Structures of dehydrocavidine (85), palmatine (86), and chelerythrine (87).

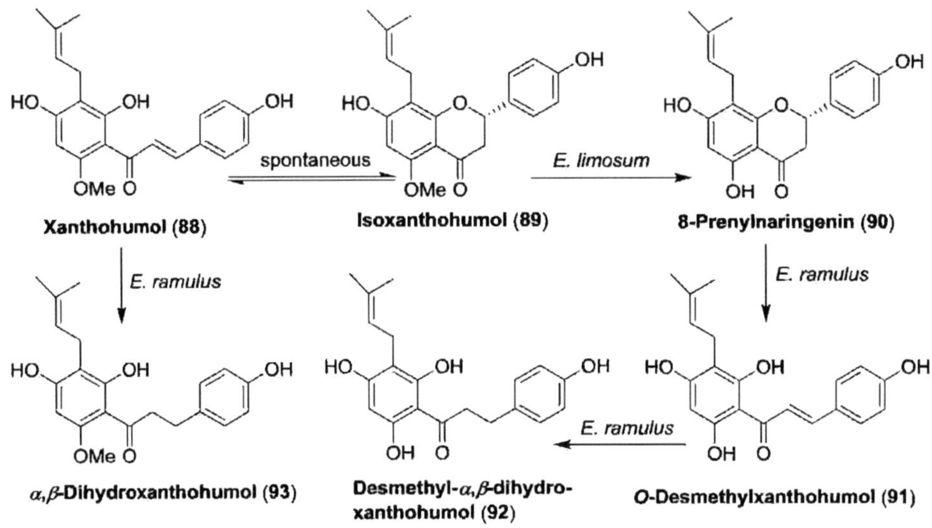

Figure 20.
Bioconversion of xanthohumol (88) to isoxanthohumol (89), 8-prenylnaringenin (90), O-desmethylxanthohumol (91), desmethyl-α,β-dihydroxanthohumol (92), and α,β-dihydroxanthohumol (93) by human gut bacteria.

able to transform xanthohumol (**88**) to *α,β*-dihydroxanthohumol (**93**) (**Figure 20**) [132]. An independent study in healthy women volunteers revealed that isoxanthohumol (**89**) could be bioconverted to 8-prenylnaringenin (**90**) in human intestine and that the bacterial microbiota isolated from fecal samples of female volunteers could also biotransform isoxanthohumol (**89**) to 8-prenylnaringenin (**90**) [133]. Another study demonstrated that 8-prenylnaringenin (**90**) has potent estrogenic property, and it could relieve climacteric symptoms, i.e., vasomotoric complaints and osteoporosis, and may be useful for the treatment of menopausal complaints [134]. These studies conclusively show that the metabolites produced by gut microbiome, i.e., 8-prenylnaringenin (**90**), are actually bioactive compounds, not the parent natural products, and they have different pharmacological activities from their parent natural products. Gut microbiome is therefore important for in vivo biotransformation of natural products, providing bioactive metabolites responsible for therapeutic effects.

Gut microbiome can biotransform natural products to bioactive metabolite essentially for therapeutic effects, for example, the biotransformation of isoxanthohumol (**89**) to bioactive 8-prenylnaringenin (**90**) [133]. However, gut microbiome can also produce toxic metabolites from the biotransformation of natural products, thus giving negative side effects. Camptothecin (CPT) is a natural alkaloid of a plant, *Camptotheca acuminata*, and has anticancer property with topoisomerase inhibitory activity [135]. Irinotecan or CPT-11 (**94**) is an alkaloid derivative of camptothecin and used as anticancer drug (**Figure 21**). Irinotecan (**94**) is a prodrug, which is transformed in vivo through hydrolysis by carboxylesterase enzymes, giving an active metabolite, SN-38 (**95**) (**Figure 21**) [136]. Uridine diphosphate-glucuronosyltransferase enzymes catalyze the conversion of SN-38 (**95**) to a glucuronidated derivative, SN-38G (**96**) (**Figure 21**). The metabolite SN-38G (**96**) is inactive for cancer cells and is excreted into the gastrointestinal tract [137], where the gut bacteria use *β*-glucuronidase enzymes to convert SN-38G (**96**) to SN-38 (**95**) that causes severe diarrhea in patients (**Figure 21**) [138]. This side effect reflects the significant negative effects of gut bacteria in the drug metabolism. However, the use of antibiotics, e.g., levofloxacin, to reduce

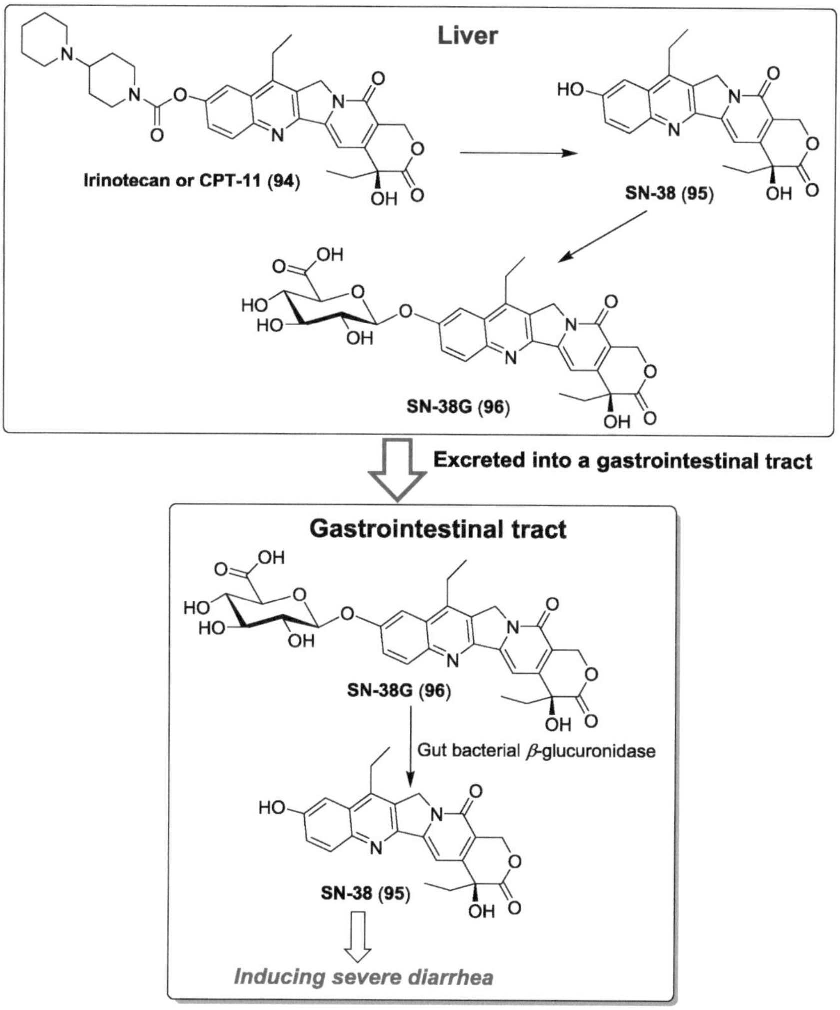

Figure 21.
*Structures of irinotecan or CPT-11 (**94**) and its metabolites, SN-38 (**95**) and SN-38G (**96**), and the bioconversion of SN-38G (**96**) to SN-38 (**95**) by gut bacterial β-glucuronidase.*

the population of gut bacteria in the gastrointestinal tract is not recommended for patients because it has many consequent problems [139]. Gut microbiotas are important for a healthy gastrointestinal tract, and they play many essential roles in dietary metabolisms [140, 141]; the treatment of cancer should not give any effects to gut microorganisms. Therefore, the use of antibiotic drugs, which affect gut microbiota, is not recommended. To reduce the diarrhea side effect without affecting gut microorganisms, a research led by Redinbo employed appropriate inhibitors of gut bacterial *β*-glucuronidase enzymes, in order to prevent the formation of SN-38 (**95**), a causative agent of severe diarrhea in patients [142]. Certain inhibitors exhibited *β*-glucuronidase inhibitory activity in living bacterial cells without disturbing the growth of gut bacteria or giving any damaging effects toward mammalian cells. Indeed, in a mouse model, mice treated with both irinotecan (**94**) and a *β*-glucuronidase inhibitor had less diarrhea and bloody diarrhea than the group receiving only the drug irinotecan (**94**). Therefore, the inhibition of microbial *β*-glucuronidases could prevent the production of toxic metabolite, SN-38 (**95**), during the treatment of anticancer drug, irinotecan (**94**) [142]. This is an example of a toxic drug metabolite produced by the activity of gut microbiome, and the manipulation of the enzyme activity of gut bacteria could be done by using another drug (an inhibitor of bacterial enzyme).

Some natural products can alter the composition of gut microbiome, and changes in gut microbiome lead to the drug-induced negative side effects on certain treatments. Paclitaxel or Taxol (**97**) is an anticancer drug for the treatment of many types of cancers (**Figure 22**), and it is a natural product isolated from a Pacific yew tree, *Taxus brevifolia*. In a mouse model, paclitaxel (**97**) chemotherapy could change the composition of gut bacterial community and induce negative effects such as sickness behaviors, i.e., fatigue and anorexia, increased central and peripheral inflammation, and impaired cognitive performance [143]. These negative effects might be associated with changes in gut bacteria because paclitaxel (**97**) therapy decreased the abundance of gut bacteria including *Lachnospiraceae* bacteria and butyrate-producing bacteria, which are necessary for human gut health [143]. Therefore, the negative effects of cancer chemotherapy may be attenuated by improving the composition of gut microbiota, for example, the use of prebiotic or probiotic supplements, which has become one of the emerging approaches to change the microbiota composition, thus improving therapeutic outcome for patients treated with anticancer drugs [144].

Antibiotic drugs have significant effects on the metabolism of drugs and phytochemicals because they could suppress enzymatic activities of gut microbiome [92]. Therefore, if patients are treated with an antibiotic drug together with another drug, there are possible drug–drug interactions due to changes of gut microbiota caused by antibiotic drugs. Lovastatin (**98**) (**Figure 23**), a natural polyketide isolated from the fungus *Aspergillus terreus* [145], is a cholesterol-lowering drug, which is a member of the statin family. Lovastatin (**98**) has the interactions with antibiotics through the mediation of gut microbiome [146]. Incubation of lovastatin (**98**) with human and rat fecalase revealed the biotransformation of this drug by gut microbiota, giving four major metabolites including demethylbutyryl-lovastatin (**99**), hydroxylated-lovastatin (**100**), hydroxy acid-lovastatin (**101**), and OH-hydroxy acid-lovastatin (**102**) (**Figure 23**) [146]. These four metabolites were also found in rat plasma, and they might be from gut microbiota-mediated metabolism of the drug lovastatin (**98**). Among the drug metabolites, hydroxy acid-lovastatin (**101**) is the active form, which could effectively inhibit 3-hydroxy-3-methylglutaryl coenzyme-A reductase, the target enzyme of this cholesterol-lowering drug [147].

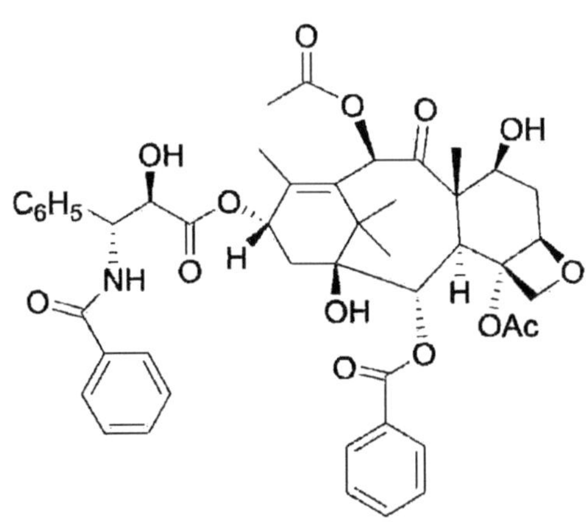

Paclitaxel or Taxol (97)

Figure 22.
Structure of an anticancer drug paclitaxel or Taxol (97).

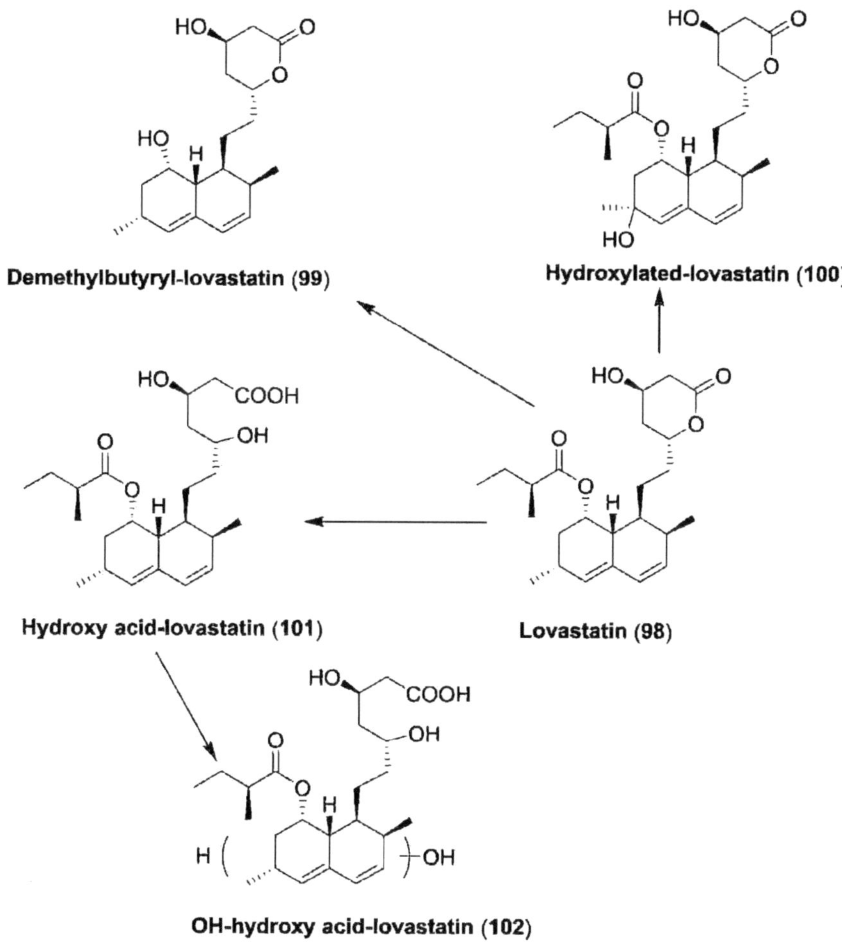

Figure 23.
Structures of major metabolites of lovastatin (98) including demethylbutyryl-lovastatin (99), hydroxylated-lovastatin (100), hydroxy acid-lovastatin (101), and OH-hydroxy acid-lovastatin (102).

In an animal model, rats with an oral administration of lovastatin (**98**) were compared with those treated with lovastatin (**98**) and antibiotics; the pharmacokinetic study revealed that the levels of the active metabolite hydroxy acid-lovastatin (**101**) in antibiotic-treated rats were lower than that without antibiotics. This result indicates that antibiotic drugs reduce the biotransformation of the drug lovastatin (**98**) to its active form, hydroxy acid-lovastatin (**101**), because antibiotics affect gut microbiome. The in vivo metabolism of lovastatin (**98**) to its active form, hydroxy acid-lovastatin (**101**), is important for the therapeutic efficacy of this drug; therefore antibiotic intake of patients treated with lovastatin (**98**) would lead to the decrease of the active form, hydroxy acid-lovastatin (**101**), thus decreasing its therapeutic effects [146]. This study clearly demonstrates the drug–drug interaction mediated by changes of gut microbiome.

4. Conclusions

Intriguingly, gut microbiome is very important for human health and diseases, and it is therefore recognized as an "organ" or a "tissue" in the human body. Gut microorganisms have much more genes encoding enzymes than those of human genome; therefore, enzymes of these microbes are involved in many biochemical processes, i.e., metabolism of xenobiotics (compounds not produced in human host, e.g., drugs and pollutants) and dietary sources. Metabolites produced by gut microbiome play significant roles in human health and diseases; these metabolites include short-chain fatty acids such as butyrate (**11**), as well as other metabolites, e.g., nicotinamide (**12**), 5-aminovaleric acid (**14**), and taurine (**15**) (see Section 2). Since gut microbiome and its metabolites substantially contribute to human health and diseases, a therapy by intervention strategies using gut microbiota can potentially be useful for some diseases, for example, metabolic disorders, cardiovascular disease, food allergy, and neurological disorders. Supplementation with probiotics or certain gut bacteria, as well as their metabolites, may be a new therapeutic method in the future. Fecal microbiota transplantation, e.g., transferring gut bacteria from healthy individuals into patients, is a challenging research study in the near future.

Gut microbiome can metabolite commonly used drugs and natural products. Drug metabolism by gut microorganisms decreases the levels of drugs in serum, thus disturbing the drug pharmacokinetics, which can lead to alteration of therapeutic efficiency. Moreover, metabolites produced by the drug metabolism of gut microbiome contribute considerably to the drug efficacy. For example, the levels of the drug L-dopa (**56**) are substantially reduced by the metabolic activity of gut microbiome, and this results in the requirement of higher doses for the Parkinson's patients with gut microbiome that has high metabolic activity toward the drug L-dopa (**56**) (see Section 3.1). This example well demonstrates the role of gut microorganisms on treatment outcomes of the commonly used drugs. Gut microbiome could improve many drug therapies, for example, cancer immunotherapy targeting CTLA-4 blockade and immune checkpoint inhibitor via the PD-1/PD-L1 pathway. Moreover, the metabolism of gut microbiome improves drug efficacy because it assists the bioconversion of some drugs into their active forms, for example, a biotransformation of lovastatin (**98**) to its active form, hydroxy acid-lovastatin (**101**), and a bioconversion of aspirin (**70**) to salicylic acid (**71**) that actively reduces pain. Interestingly, gut microbiome involves in a biotransformation of an alkaloid natural product berberine (**78**) to an absorbable form, dihydroberberine (**80**), which is absorbed at the intestine system (see Section 3.2). This result demonstrates that gut microbiome facilitates drug delivery of berberine (**78**) that has poor solubility by a biotransformation to an absorbable form, dihydroberberine (**80**), which is in turn converted to its active form berberine (**78**) in the human body. Since gut microbiome plays many important roles in drugs and natural products, the metabolism of natural products and drug candidates by gut microbiome should therefore be studied, and it should be a part of the drug development process. Gut microbiome can potentially play a crucial role for the improvement of drug safety and efficacy.

Acknowledgements

The author thanks the Center of Excellence on Environmental Health and Toxicology, Science & Technology Postgraduate Education and Research Development Office (PERDO), Ministry of Education, for the support of research that leads to this book chapter.

Conflict of interest

The author declares no competing interests or no conflict of interest.

Author details

Prasat Kittakoop[1,2,3]

1 Chulabhorn Graduate Institute, Program in Chemical Sciences, Chulabhorn Royal Academy, Bangkok, Thailand

2 Chulabhorn Research Institute, Bangkok, Thailand

3 Center of Excellence on Environmental Health and Toxicology (EHT), CHE, Ministry of Education, Thailand

*Address all correspondence to: prasat@cri.or.th

References

[1] Sanders ME, Merenstein DJ, Reid G, Gibson GR, Rastall RA. Probiotics and prebiotics in intestinal health and disease: From biology to the clinic. Nature Reviews. Gastroenterology & Hepatology. 2019;**16**(10):605-616. DOI: 10.1038/s41575-019-0173-3

[2] Matthies A, Loh G, Blaut M, Braune A. Daidzein and genistein are converted to equol and 5-hydroxy-equol by human intestinal Slackia isoflavoniconvertens in gnotobiotic rats. The Journal of Nutrition. 2012;**142**(1):40-46. DOI: 10.3945/jn.111.148247

[3] Matthies A, Clavel T, Gutschow M, Engst W, Haller D, Blaut M, et al. Conversion of daidzein and genistein by an anaerobic bacterium newly isolated from the mouse intestine. Applied and Environmental Microbiology. 2008;**74**(15):4847-4852. DOI: 10.1128/AEM.00555-08

[4] Setchell KD, Clerici C, Lephart ED, Cole SJ, Heenan C, Castellani D, et al. *S*-equol, a potent ligand for estrogen receptor beta, is the exclusive enantiomeric form of the soy isoflavone metabolite produced by human intestinal bacterial flora. The American Journal of Clinical Nutrition. 2005;**81**(5):1072-1079. DOI: 10.1093/ajcn/81.5.1072

[5] Raimondi S, Roncaglia L, De Lucia M, Amaretti A, Leonardi A, Pagnoni UM, et al. Bioconversion of soy isoflavones daidzin and daidzein by *Bifidobacterium* strains. Applied Microbiology and Biotechnology. 2009;**81**(5):943-950. DOI: 10.1007/s00253-008-1719-4

[6] Frankenfeld CL. *O*-Desmethylangolensin: The importance of equol's lesser known cousin to human health. Advances in Nutrition. 2011;**2**(4):317-324. DOI: 10.3945/an.111.000539

[7] Rowland I, Wiseman H, Sanders T, Adlercreutz H, Bowey E. Metabolism of oestrogens and phytoestrogens: Role of the gut microflora. Biochemical Society Transactions. 1999;**27**(2):304-308. DOI: 10.1042/bst0270304

[8] Fujioka M, Uehara M, Wu J, Adlercreutz H, Suzuki K, Kanazawa K, et al. Equol, a metabolite of daidzein, inhibits bone loss in ovariectomized mice. The Journal of Nutrition. 2004;**134**(10):2623-2627. DOI: 10.1093/jn/134.10.2623

[9] Kimira Y, Katsumata S, Suzuki K, Ishimi Y, Ueno T, Uchiyama S, et al. Comparative activities of the *S*-enantiomer and racemic forms of equol on bone fragility in ovariectomized mice. Bioscience, Biotechnology, and Biochemistry. 2012;**76**(5):1018-1021. DOI: 10.1271/bbb.110973

[10] Shimada Y, Takahashi M, Miyazawa N, Abiru Y, Uchiyama S, Hishigaki H. Identification of a novel dihydrodaidzein racemase essential for biosynthesis of equol from daidzein in *Lactococcus* sp. strain 20-92. Applied and Environmental Microbiology. 2012;**78**(14):4902-4907. DOI: 10.1128/AEM.00410-12

[11] Burcelin R. Regulation of metabolism: A cross talk between gut microbiota and its human host. Physiology. 2012;**27**(5):300-307. DOI: 10.1152/physiol.00023.2012

[12] O'Hara AM, Shanahan F. The gut flora as a forgotten organ. EMBO Reports. 2006;**7**(7):688-693. DOI: 10.1038/sj.embor.7400731

[13] Zhu B, Wang X, Li L. Human gut microbiome: The second genome of human body. Protein & Cell. 2010;**1**(8):718-725. DOI: 10.1007/s13238-010-0093-z

[14] Iacob S, Iacob DG. Infectious threats, the intestinal barrier, and its Trojan horse: Dysbiosis. Frontiers in Microbiology. 2019;**10**:1676. DOI: 10.3389/fmicb.2019.01676

[15] Hills RD Jr, Pontefract BA, Mishcon HR, Black CA, Sutton SC, Theberge CR. Gut microbiome: Profound implications for diet and disease. Nutrients. 2019;**11**(7):1613. DOI: 10.3390/nu11071613

[16] Carding S, Verbeke K, Vipond DT, Corfe BM, Owen LJ. Dysbiosis of the gut microbiota in disease. Microbial Ecology in Health and Disease. 2015;**26**:26191. DOI: 10.3402/mehd.v26.26191

[17] Rooks MG, Garrett WS. Gut microbiota, metabolites and host immunity. Nature Reviews. Immunology. 2016;**16**(6):341-352. DOI: 10.1038/nri.2016.42

[18] Li Z, Quan G, Jiang X, Yang Y, Ding X, Zhang D, et al. Effects of metabolites derived from gut microbiota and hosts on pathogens. Frontiers in Cellular and Infection Microbiology. 2018;**8**:314. DOI: 10.3389/fcimb.2018.00314

[19] Louis P, Hold GL, Flint HJ. The gut microbiota, bacterial metabolites and colorectal cancer. Nature Reviews. Microbiology. 2014;**12**(10):661-672. DOI: 10.1038/nrmicro3344

[20] Caspani G, Kennedy S, Foster JA, Swann J. Gut microbial metabolites in depression: Understanding the biochemical mechanisms. Microbial Cell. 2019;**6**(10):454-481. DOI: 10.15698/mic2019.10.693

[21] Zeng H, Umar S, Rust B, Lazarova D, Bordonaro M. Secondary bile acids and short chain fatty acids in the colon: A focus on colonic microbiome, cell proliferation, inflammation, and cancer. International Journal of Molecular Sciences. 2019;**20**(5):1214. DOI: 10.3390/ijms20051214

[22] Brial F, Le Lay A, Dumas ME, Gauguier D. Implication of gut microbiota metabolites in cardiovascular and metabolic diseases. Cellular and Molecular Life Sciences. 2018;**75**(21):3977-3990. DOI: 10.1007/s00018-018-2901-1

[23] Wang Z, Zhao Y. Gut microbiota derived metabolites in cardiovascular health and disease. Protein & Cell. 2018;**9**(5):416-431. DOI: 10.1007/s13238-018-0549-0

[24] den Besten G, van Eunen K, Groen AK, Venema K, Reijngoud DJ, Bakker BM. The role of short-chain fatty acids in the interplay between diet, gut microbiota, and host energy metabolism. Journal of Lipid Research. 2013;**54**(9):2325-2340. DOI: 10.1194/jlr.R036012

[25] Furusawa Y, Obata Y, Fukuda S, Endo TA, Nakato G, Takahashi D, et al. Commensal microbe-derived butyrate induces the differentiation of colonic regulatory T cells. Nature. 2013;**504**(7480):446-450. DOI: 10.1038/nature12721

[26] Felizardo RJF, Watanabe IKM, Dardi P, Rossoni LV, Camara NOS. The interplay among gut microbiota, hypertension and kidney diseases: The role of short-chain fatty acids. Pharmacological Research. 2019;**141**:366-377. DOI: 10.1016/j.phrs.2019.01.019

[27] Parada Venegas D, De la Fuente MK, Landskron G, Gonzalez MJ, Quera R, Dijkstra G, et al. Short chain fatty acids (SCFAs)-mediated gut epithelial and immune regulation and its relevance for inflammatory bowel diseases. Frontiers in Immunology. 2019;**10**:277. DOI: 10.3389/fimmu.2019.00277

[28] Zhuang X, Li T, Li M, Huang S, Qiu Y, Feng R, et al. Systematic

review and meta-analysis: Short-chain fatty acid characterization in patients with inflammatory bowel disease. Inflammatory Bowel Diseases. 2019;**25**(11):1751-1763. DOI: 10.1093/ibd/izz188

[29] Yang G, Chen S, Deng B, Tan C, Deng J, Zhu G, et al. Implication of G protein-coupled receptor 43 in intestinal inflammation: A mini-review. Frontiers in Immunology. 2018;**9**:1434. DOI: 10.3389/fimmu.2018.01434

[30] Koh A, De Vadder F, Kovatcheva-Datchary P, Backhed F. From dietary fiber to host physiology: Short-chain fatty acids as key bacterial metabolites. Cell. 2016;**165**(6):1332-1345. DOI: 10.1016/j.cell.2016.05.041

[31] Fellows R, Denizot J, Stellato C, Cuomo A, Jain P, Stoyanova E, et al. Microbiota derived short chain fatty acids promote histone crotonylation in the colon through histone deacetylases. Nature Communications. 2018;**9**(1):105. DOI: 10.1038/s41467-017-02651-5

[32] Sanna S, van Zuydam NR, Mahajan A, Kurilshikov A, Vich Vila A, Vosa U, et al. Causal relationships among the gut microbiome, short-chain fatty acids and metabolic diseases. Nature Genetics. 2019;**51**(4):600-605. DOI: 10.1038/s41588-019-0350-x

[33] Larsen N, Vogensen FK, van den Berg FW, Nielsen DS, Andreasen AS, Pedersen BK, et al. Gut microbiota in human adults with type 2 diabetes differs from non-diabetic adults. PLoS One. 2010;**5**(2):e9085. DOI: 10.1371/journal.pone.0009085

[34] Hu J, Lin S, Zheng B, Cheung PCK. Short-chain fatty acids in control of energy metabolism. Critical Reviews in Food Science and Nutrition. 2018;**58**(8):1243-1249. DOI: 10.1080/10408398.2016.1245650

[35] Zhao W, Ho HE, Bunyavanich S. The gut microbiome in food allergy. Annals of Allergy, Asthma & Immunology. 2019;**122**(3):276-282. DOI: 10.1016/j.anai.2018.12.012

[36] Berni Canani R, Paparo L, Nocerino R, Di Scala C, Della Gatta G, Maddalena Y, et al. Gut microbiome as target for innovative strategies against food allergy. Frontiers in Immunology. 2019;**10**:191. DOI: 10.3389/fimmu.2019.00191

[37] Rachid R, Chatila TA. The role of the gut microbiota in food allergy. Current Opinion in Pediatrics. 2016;**28**(6):748-753. DOI: 10.1097/MOP.0000000000000427

[38] Feehley T, Plunkett CH, Bao R, Choi Hong SM, Culleen E, Belda-Ferre P, et al. Healthy infants harbor intestinal bacteria that protect against food allergy. Nature Medicine. 2019;**25**(3):448-453. DOI: 10.1038/s41591-018-0324-z

[39] Schwiertz A, Hold GL, Duncan SH, Gruhl B, Collins MD, Lawson PA, et al. *Anaerostipes caccae* gen. nov., sp. nov., a new saccharolytic, acetate-utilising, butyrate-producing bacterium from human faeces. Systematic and Applied Microbiology. 2002;**25**(1):46-51. DOI: 10.1078/0723-2020-00096

[40] Donohoe DR, Garge N, Zhang X, Sun W, O'Connell TM, Bunger MK, et al. The microbiome and butyrate regulate energy metabolism and autophagy in the mammalian colon. Cell Metabolism. 2011;**13**(5):517-526. DOI: 10.1016/j.cmet.2011.02.018

[41] Berni Canani R, Sangwan N, Stefka AT, Nocerino R, Paparo L, Aitoro R, et al. *Lactobacillus rhamnosus* GG-supplemented formula expands butyrate-producing bacterial strains in food allergic infants. The ISME Journal. 2016;**10**(3):742-750. DOI: 10.1038/ismej.2015.151

[42] Fricker RA, Green EL, Jenkins SI, Griffin SM. The influence of nicotinamide on health and disease

in the central nervous system. International Journal of Tryptophan Research. 2018;**11**:1-11. DOI: 10.1177/1178646918776658

[43] Blacher E, Bashiardes S, Shapiro H, Rothschild D, Mor U, Dori-Bachash M, et al. Potential roles of gut microbiome and metabolites in modulating ALS in mice. Nature. 2019;**572**(7770):474-480. DOI: 10.1038/s41586-019-1443-5

[44] Depommier C, Everard A, Druart C, Plovier H, Van Hul M, Vieira-Silva S, et al. Supplementation with *Akkermansia muciniphila* in overweight and obese human volunteers: A proof-of-concept exploratory study. Nature Medicine. 2019;**25**(7):1096-1103. DOI: 10.1038/s41591-019-0495-2

[45] Xu M, Xu X, Li J, Li F. Association between gut microbiota and autism spectrum disorder: A systematic review and meta-analysis. Frontiers in Psychiatry. 2019;**10**:473. DOI: 10.3389/fpsyt.2019.00473

[46] Sharon G, Cruz NJ, Kang DW, Gandal MJ, Wang B, Kim YM, et al. Human gut microbiota from autism spectrum disorder promote behavioral symptoms in mice. Cell. 2019;**177**(6):1600-1618.e17. DOI: 10.1016/j.cell.2019.05.004

[47] Callery PS, Geelhaar LA. 1-Piperideine as an in vivo precursor of the gamma-aminobutyric acid homologue 5-aminopentanoic acid. Journal of Neurochemistry. 1985;**45**(3):946-968. DOI: 10.1111/j.1471-4159.1985.tb04085.x

[48] Park E, Cohen I, Gonzalez M, Castellano MR, Flory M, Jenkins EC, et al. Is taurine a biomarker in autistic spectrum disorder? Advances in Experimental Medicine and Biology. 2017;**975**(Pt 1):3-16. DOI: 10.1007/978-94-024-1079-2_1

[49] Seo DO, Holtzman DM. Gut microbiota: From the forgotten organ to a potential key player in the pathology of Alzheimer disease. The Journals of Gerontology: Series A. 2019:glz262. DOI: 10.1093/gerona/glz262

[50] Clarke G, Stilling RM, Kennedy PJ, Stanton C, Cryan JF, Dinan TG. Minireview: Gut microbiota: The neglected endocrine organ. Molecular Endocrinology. 2014;**28**(8):1221-1238. DOI: 10.1210/me.2014-1108

[51] Guinane CM, Cotter PD. Role of the gut microbiota in health and chronic gastrointestinal disease: Understanding a hidden metabolic organ. Therapeutic Advances in Gastroenterology. 2013;**6**(4):295-308. DOI: 10.1177/1756283X13482996

[52] Li X, Liu L, Cao Z, Li W, Li H, Lu C, et al. Gut microbiota as an "invisible organ" that modulates the function of drugs. Biomedicine & Pharmacotherapy. 2020;**121**:109653. DOI: 10.1016/j.biopha.2019.109653

[53] Lubomski M, Davis RL, Sue CM. The gut microbiota: A novel therapeutic target in Parkinson's disease? Parkinsonism & Related Disorders. 2019;**66**:265-266. DOI: 10.1016/j.parkreldis.2019.08.010

[54] Yang S, Li X, Yang F, Zhao R, Pan X, Liang J, et al. Gut microbiota-dependent marker TMAO in promoting cardiovascular disease: Inflammation mechanism, clinical prognostic, and potential as a therapeutic target. Frontiers in Pharmacology. 2019;**10**:1360. DOI: 10.3389/fphar.2019.01360

[55] Okubo H, Nakatsu Y, Kushiyama A, Yamamotoya T, Matsunaga Y, Inoue MK, et al. Gut microbiota as a therapeutic target for metabolic disorders. Current Medicinal Chemistry. 2018;**25**(9):984-1001. DOI: 10.2174/0929867324666171009121702

[56] Amedei A. Potential therapeutic strategies to target gut microbiota in

hepatocellular carcinoma. Hepatobiliary Surgery and Nutrition. 2019;**8**(5):527-529. DOI: 10.21037/hbsn.2019.09.04

[57] Suk KT, Kim DJ. Gut microbiota: Novel therapeutic target for nonalcoholic fatty liver disease. Expert Review of Gastroenterology & Hepatology. 2019;**13**(3):193-204. DOI: 10.1080/17474124.2019.1569513

[58] Aitoro R, Paparo L, Amoroso A, Di Costanzo M, Cosenza L, Granata V, et al. Gut microbiota as a target for preventive and therapeutic intervention against food allergy. Nutrients. 2017;**9**(7):672. DOI: 10.3390/nu9070672

[59] Zhao S, Gao G, Li W, Li X, Zhao C, Jiang T, et al. Antibiotics are associated with attenuated efficacy of anti-PD-1/PD-L1 therapies in Chinese patients with advanced non-small cell lung cancer. Lung Cancer. 2019;**130**:10-17. DOI: 10.1016/j.lungcan.2019.01.017

[60] Sender R, Fuchs S, Milo R. Revised estimates for the number of human and bacteria cells in the body. PLoS Biology. 2016;**14**(8):e1002533. DOI: 10.1371/journal.pbio.1002533

[61] Qin J, Li R, Raes J, Arumugam M, Burgdorf KS, Manichanh C, et al. A human gut microbial gene catalogue established by metagenomic sequencing. Nature. 2010;**464**(7285):59-65. DOI: 10.1038/nature08821

[62] Zimmermann M, Zimmermann-Kogadeeva M, Wegmann R, Goodman AL. Mapping human microbiome drug metabolism by gut bacteria and their genes. Nature. 2019;**570**(7762):462-467. DOI: 10.1038/s41586-019-1291-3

[63] Ianiro G, Tilg H, Gasbarrini A. Antibiotics as deep modulators of gut microbiota: Between good and evil. Gut. 2016;**65**(11):1906-1915. DOI: 10.1136/gutjnl-2016-312297

[64] Kim S, Covington A, Pamer EG. The intestinal microbiota: Antibiotics, colonization resistance, and enteric pathogens. Immunological Reviews. 2017;**279**(1):90-105. DOI: 10.1111/imr.12563

[65] Maier L, Pruteanu M, Kuhn M, Zeller G, Telzerow A, Anderson EE, et al. Extensive impact of non-antibiotic drugs on human gut bacteria. Nature. 2018;**555**(7698):623-628. DOI: 10.1038/nature25979

[66] Zimmermann M, Zimmermann-Kogadeeva M, Wegmann R, Goodman AL. Separating host and microbiome contributions to drug pharmacokinetics and toxicity. Science. 2019;**363**(6427):eaat9931. DOI: 10.1126/science.aat9931

[67] Nakayama H, Kinouchi T, Kataoka K, Akimoto S, Matsuda Y, Ohnishi Y. Intestinal anaerobic bacteria hydrolyse sorivudine, producing the high blood concentration of 5-(E)-(2-bromovinyl)uracil that increases the level and toxicity of 5-fluorouracil. Pharmacogenetics. 1997;7(1):35-43. DOI: 10.1097/00008571-199702000-00005

[68] Elmer GW, Remmel RP. Role of the intestinal microflora in clonazepam metabolism in the rat. Xenobiotica. 1984;**14**(11):829-840. DOI: 10.3109/00498258409151481

[69] Wilson ID, Nicholson JK. Gut microbiome interactions with drug metabolism, efficacy, and toxicity. Translational Research. 2017;**179**:204-222. DOI: 10.1016/j.trsl.2016.08.002

[70] Winter J, Morris GN, O'Rourke-Locascio S, Bokkenheuser VD, Mosbach EH, Cohen BI, et al. Mode of action of steroid desmolase and reductases synthesized by *Clostridium "scindens"* (formerly *Clostridium* strain 19). Journal of Lipid Research. 1984;**25**(10):1124-1131. Available from: https://www.jlr.org/content/25/10/1124.long

[71] Morris GN, Winter J, Cato EP, Ritchie AE, Bokkenheuser VD. *Clostridium scindens* sp. nov., a human intestinal bacterium with desmolytic activity on corticoids. International Journal of Systematic and Evolutionary Microbiology. 1985;**35**(4):478-481. DOI: 10.1099/00207713-35-4-478

[72] Yeung PK, Prescott C, Haddad C, Montague TJ, McGregor C, Quilliam MA, et al. Pharmacokinetics and metabolism of diltiazem in healthy males and females following a single oral dose. European Journal of Drug Metabolism and Pharmacokinetics. 1993;**18**(2):199-206. DOI: 10.1007/BF03188796

[73] Goldin BR, Peppercorn MA, Goldman P. Contributions of host and intestinal microflora in the metabolism of L-dopa by the rat. The Journal of Pharmacology and Experimental Therapeutics. 1973;**186**(1):160-166. Available from: http://jpet.aspetjournals.org/content/186/1/160.long

[74] Sandler M, Goodwin BL, Ruthven CR. Therapeutic implications in Parkinsonism of *m*-tyramine formation from L-dopa in man. Nature. 1971;**229**(5284):414-415. DOI: 10.1038/229414a0

[75] Goldman P, Peppercorn MA, Goldin BR. Metabolism of drugs by microorganisms in the intestine. The American Journal of Clinical Nutrition. 1974;**27**(11):1348-1355. DOI: 10.1093/ajcn/27.11.1348

[76] Fasano A, Bove F, Gabrielli M, Petracca M, Zocco MA, Ragazzoni E, et al. The role of small intestinal bacterial overgrowth in Parkinson's disease. Movement Disorders. 2013;**28**(9):1241-1249. DOI: 10.1002/mds.25522

[77] Maini Rekdal V, Bess EN, Bisanz JE, Turnbaugh PJ, Balskus EP. Discovery and inhibition of an interspecies gut bacterial pathway for Levodopa metabolism. Science. 2019;**364**(6445):eaau6323. DOI: 10.1126/science.aau6323

[78] van Kessel SP, Frye AK, El-Gendy AO, Castejon M, Keshavarzian A, van Dijk G, et al. Gut bacterial tyrosine decarboxylases restrict levels of levodopa in the treatment of Parkinson's disease. Nature Communications. 2019;**10**(1):310. DOI: 10.1038/s41467-019-08294-y

[79] Vetizou M, Pitt JM, Daillere R, Lepage P, Waldschmitt N, Flament C, et al. Anticancer immunotherapy by CTLA-4 blockade relies on the gut microbiota. Science. 2015;**350**(6264):1079-1084. DOI: 10.1126/science.aad1329

[80] Matson V, Fessler J, Bao R, Chongsuwat T, Zha Y, Alegre M-L, et al. The commensal microbiome is associated with anti-PD-1 efficacy in metastatic melanoma patients. Science. 2018;**359**(6371):104-108. DOI: 10.1126/science.aao3290

[81] Routy B, Le Chatelier E, Derosa L, Duong CPM, Alou MT, Daillère R, et al. Gut microbiome influences efficacy of PD-1-based immunotherapy against epithelial tumors. Science. 2018;**359**(6371):91-97. DOI: 10.1126/science.aan3706

[82] Gopalakrishnan V, Spencer CN, Nezi L, Reuben A, Andrews MC, Karpinets TV, et al. Gut microbiome modulates response to anti-PD-1 immunotherapy in melanoma patients. Science. 2018;**359**(6371):97-103. DOI: 10.1126/science.aan4236

[83] Ma W, Mao Q, Xia W, Dong G, Yu C, Jiang F. Gut microbiota shapes the efficiency of cancer therapy. Frontiers in Microbiology. 2019;**10**:1050. DOI: 10.3389/fmicb.2019.01050

[84] Viaud S, Saccheri F, Mignot G, Yamazaki T, Daillere R, Hannani D,

et al. The intestinal microbiota modulates the anticancer immune effects of cyclophosphamide. Science. 2013;**342**(6161):971-976. DOI: 10.1126/science.1240537

[85] Daillere R, Vetizou M, Waldschmitt N, Yamazaki T, Isnard C, Poirier-Colame V, et al. *Enterococcus hirae* and *Barnesiella intestinihominis* facilitate cyclophosphamide-induced therapeutic immunomodulatory effects. Immunity. 2016;**45**(4):931-943. DOI: 10.1016/j.immuni.2016.09.009

[86] Geller LT, Barzily-Rokni M, Danino T, Jonas OH, Shental N, Nejman D, et al. Potential role of intratumor bacteria in mediating tumor resistance to the chemotherapeutic drug gemcitabine. Science. 2017;**357**(6356):1156-1160. DOI: 10.1126/science.aah5043

[87] Plunkett W, Huang P, Xu YZ, Heinemann V, Grunewald R, Gandhi V. Gemcitabine: Metabolism, mechanisms of action, and self-potentiation. Seminars in Oncology. 1995;**22**(4 Suppl 11):3-10. Available from: https://scinapse.io/papers/2284799106

[88] Vande Voorde J, Sabuncuoglu S, Noppen S, Hofer A, Ranjbarian F, Fieuws S, et al. Nucleoside-catabolizing enzymes in mycoplasma-infected tumor cell cultures compromise the cytostatic activity of the anticancer drug gemcitabine. The Journal of Biological Chemistry. 2014;**289**(19):13054-13065. DOI: 10.1074/jbc.M114.558924

[89] Yu T, Guo F, Yu Y, Sun T, Ma D, Han J, et al. *Fusobacterium nucleatum* promotes chemoresistance to colorectal cancer by modulating autophagy. Cell. 2017;**170**(3):548-563.e16. DOI: 10.1016/j.cell.2017.07.008

[90] Ganesan K, Guo S, Fayyaz S, Zhang G, Xu B. Targeting programmed *Fusobacterium nucleatum* Fap2 for colorectal cancer therapy. Cancers. 2019;**11**(10):1592. DOI: 10.3390/cancers11101592

[91] Gethings-Behncke C, Coleman HG, Jordao HWT, Longley DB, Crawford N, Murray LJ, et al. *Fusobacterium nucleatum* in the colorectum and its association with cancer risk and survival: A systematic review and meta-analysis. Cancer Epidemiology, Biomarkers & Prevention. 2020;**29**(3):539-548. DOI: 10.1158/1055-9965.EPI-18-1295

[92] Kim D-H. Gut microbiota-mediated drug-antibiotic interactions. Drug Metabolism and Disposition. 2015;**43**(10):1581-1589. DOI: 10.1124/dmd.115.063867

[93] Zhang J, Sun Y, Wang R, Zhang J. Gut microbiota-mediated drug-drug interaction between amoxicillin and aspirin. Scientific Reports. 2019;**9**(1):16194. DOI: 10.1038/s41598-019-52632-5

[94] Patrono C, Baigent C. Role of aspirin in primary prevention of cardiovascular disease. Nature Reviews. Cardiology. 2019;**16**(11):675-686. DOI: 10.1038/s41569-019-0225-y

[95] Zhang X, Feng Y, Liu X, Ma J, Li Y, Wang T, et al. Beyond a chemopreventive reagent, aspirin is a master regulator of the hallmarks of cancer. Journal of Cancer Research and Clinical Oncology. 2019;**145**(6):1387-1403. DOI: 10.1007/s00432-019-02902-6

[96] Choi HW, Tian M, Song F, Venereau E, Preti A, Park S-W, et al. Aspirin's active metabolite salicylic acid targets high mobility group box 1 to modulate inflammatory responses. Molecular Medicine. 2015;**21**(1):526-535. DOI: 10.2119/molmed.2015.00148

[97] Kim IS, Yoo DH, Jung IH, Lim S, Jeong JJ, Kim KA, et al. Reduced

metabolic activity of gut microbiota by antibiotics can potentiate the antithrombotic effect of aspirin. Biochemical Pharmacology. 2016;**122**:72-79. DOI: 10.1016/j.bcp.2016.09.023

[98] Sun Y, Zhang J, Zhao A, Li W, Feng Q, Wang R. Effects of intestinal flora on the pharmacokinetics and pharmacodynamics of aspirin in high-altitude hypoxia. PLoS One. 2020;**15**(3):e0230197. DOI: 10.1371/journal.pone.0230197

[99] Zhang J, Chen Y, Sun Y, Wang R, Zhang J, Jia Z. Plateau hypoxia attenuates the metabolic activity of intestinal flora to enhance the bioavailability of nifedipine. Drug Delivery. 2018;**25**(1):1175-1181. DOI: 10.1080/10717544.2018.1469687

[100] Poolchanuan P, Unagul P, Thongnest S, Wiyakrutta S, Ngamrojanavanich N, Mahidol C, et al. An anticonvulsive drug, valproic acid (valproate), has effects on the biosynthesis of fatty acids and polyketides in microorganisms. Scientific Reports. 2020;**10**:9300. DOI: 10.1038/s41598-020-66251-y

[101] Phiel CJ, Zhang F, Huang EY, Guenther MG, Lazar MA, Klein PS. Histone deacetylase is a direct target of valproic acid, a potent anticonvulsant, mood stabilizer, and teratogen. The Journal of Biological Chemistry. 2001;**276**(39):36734-36741. DOI: 10.1074/jbc.M101287200

[102] Vaughan EE, Heilig HG, Ben-Amor K, de Vos WM. Diversity, vitality and activities of intestinal lactic acid bacteria and bifidobacteria assessed by molecular approaches. FEMS Microbiology Reviews. 2005;**29**(3):477-490. DOI: 10.1016/j.femsre.2005.04.009

[103] Chehoud C, Albenberg LG, Judge C, Hoffmann C, Grunberg S, Bittinger K, et al. Fungal signature in the gut microbiota of pediatric patients with inflammatory bowel disease. Inflammatory Bowel Diseases. 2015;**21**(8):1948-1956. DOI: 10.1097/MIB.0000000000000454

[104] Nash AK, Auchtung TA, Wong MC, Smith DP, Gesell JR, Ross MC, et al. The gut mycobiome of the human microbiome project healthy cohort. Microbiome. 2017;**5**(1):153. DOI: 10.1186/s40168-017-0373-4

[105] Zacherl JR, Mihalik SJ, Chace DH, Christensen TC, Robinson LJ, Blair HC. Elaidate, an 18-carbon *trans*-monoenoic fatty acid, inhibits beta-oxidation in human peripheral blood macrophages. Journal of Cellular Biochemistry. 2014;**115**(1):62-70. DOI: 10.1002/jcb.24633

[106] Zacherl JR, Tourkova I, St Croix CM, Robinson LJ, Peck Palmer OM, Mihalik SJ, et al. Elaidate, an 18-carbon *trans*-monoenoic fatty acid, but not physiological fatty acids increases intracellular $Zn^{(2+)}$ in human macrophages. Journal of Cellular Biochemistry. 2015;**116**(4):524-532. DOI: 10.1002/jcb.25002

[107] Smith S, Tsai SC. The type I fatty acid and polyketide synthases: A tale of two megasynthases. Natural Product Reports. 2007;**24**(5):1041-1072. DOI: 10.1039/b603600g

[108] Wang L, Ravichandran V, Yin Y, Yin J, Zhang Y. Natural products from mammalian gut microbiota. Trends in Biotechnology. 2019;**37**(5):492-504. DOI: 10.1016/j.tibtech.2018.10.003

[109] Li BY, Xu XY, Gan RY, Sun QC, Meng JM, Shang A, et al. Targeting gut microbiota for the prevention and management of diabetes mellitus by dietary natural products. Food. 2019;**8**(10):440. DOI: 10.3390/foods8100440

[110] Tong X, Xu J, Lian F, Yu X, Zhao Y, Xu L, et al. Structural

alteration of gut microbiota during the amelioration of human type 2 diabetes with hyperlipidemia by metformin and a traditional Chinese herbal formula: a multicenter, randomized, open label clinical trial. mBio. 2018;**9**(3):e02392-e02317. DOI: 10.1128/mBio.02392-17

[111] Zhang X, Zhao Y, Zhang M, Pang X, Xu J, Kang C, et al. Structural changes of gut microbiota during berberine-mediated prevention of obesity and insulin resistance in high-fat diet-fed rats. PLoS One. 2012;**7**(8):e42529. DOI: 10.1371/journal.pone.0042529

[112] Zhang X, Zhao Y, Xu J, Xue Z, Zhang M, Pang X, et al. Modulation of gut microbiota by berberine and metformin during the treatment of high-fat diet-induced obesity in rats. Scientific Reports. 2015;**5**(1):14405. DOI: 10.1038/srep14405

[113] Cui H, Cai Y, Wang L, Jia B, Li J, Zhao S, et al. Berberine regulates Treg/Th17 balance to treat ulcerative colitis through modulating the gut microbiota in the colon. Frontiers in Pharmacology. 2018;**9**:571. DOI: 10.3389/fphar.2018.00571

[114] Massot-Cladera M, Perez-Berezo T, Franch A, Castell M, Perez-Cano FJ. Cocoa modulatory effect on rat faecal microbiota and colonic crosstalk. Archives of Biochemistry and Biophysics. 2012;**527**(2):105-112. DOI: 10.1016/j.abb.2012.05.015

[115] Martin-Pelaez S, Camps-Bossacoma M, Massot-Cladera M, Rigo-Adrover M, Franch A, Perez-Cano FJ, et al. Effect of cocoa's theobromine on intestinal microbiota of rats. Molecular Nutrition & Food Research. 2017;**61**(10):1700238. DOI: 10.1002/mnfr.201700238

[116] Palm NW, de Zoete MR, Cullen TW, Barry NA, Stefanowski J, Hao L, et al. Immunoglobulin A coating identifies colitogenic bacteria in inflammatory bowel disease. Cell. 2014;**158**(5):1000-1010. DOI: 10.1016/j.cell.2014.08.006

[117] Feng R, Shou J-W, Zhao Z-X, He C-Y, Ma C, Huang M, et al. Transforming berberine into its intestine-absorbable form by the gut microbiota. Scientific Reports. 2015;**5**(1):12155. DOI: 10.1038/srep12155

[118] Turner N, Li JY, Gosby A, To SWC, Cheng Z, Miyoshi H, et al. Berberine and its more biologically available derivative, dihydroberberine, inhibit mitochondrial respiratory complex I: A mechanism for the action of berberine to activate AMP-activated protein kinase and improve insulin action. Diabetes. 2008;**57**(5):1414-1418. DOI: 10.2337/db07-1552

[119] Tan L, Wang Y, Ai G, Luo C, Chen H, Li C, et al. Dihydroberberine, a hydrogenated derivative of berberine firstly identified in Phellodendri Chinese Cortex, exerts anti-inflammatory effect via dual modulation of NF-kappaB and MAPK signaling pathways. International Immunopharmacology. 2019;**75**:105802. DOI: 10.1016/j.intimp.2019.105802

[120] Dai B, Ma Y, Wang W, Zhan Y, Zhang D, Liu R, et al. Dihydroberberine exhibits synergistic effects with sunitinib on NSCLC NCI-H460 cells by repressing MAP kinase pathways and inflammatory mediators. Journal of Cellular and Molecular Medicine. 2017;**21**(10):2573-2585. DOI: 10.1111/jcmm.13178

[121] Yu D, Lv L, Fang L, Zhang B, Wang J, Zhan G, et al. Inhibitory effects and mechanism of dihydroberberine on hERG channels expressed in HEK293 cells. PLoS One. 2017;**12**(8):e0181823. DOI: 10.1371/journal.pone.0181823

[122] Zuo F, Nakamura N, Akao T, Hattori M. Pharmacokinetics of

berberine and its main metabolites in conventional and pseudo germ-free rats determined by liquid chromatography/ion trap mass spectrometry. Drug Metabolism and Disposition. 2006;**34**(12):2064-2072. DOI: 10.1124/dmd.106.011361

[123] Zhang P, Qiang X, Zhang M, Ma D, Zhao Z, Zhou C, et al. Demethyleneberberine, a natural mitochondria-targeted antioxidant, inhibits mitochondrial dysfunction, oxidative stress, and steatosis in alcoholic liver disease mouse model. The Journal of Pharmacology and Experimental Therapeutics. 2015;**352**(1):139-147. DOI: 10.1124/jpet.114.219832

[124] Yu XT, Xu YF, Huang YF, Qu C, Xu LQ, Su ZR, et al. Berberrubine attenuates mucosal lesions and inflammation in dextran sodium sulfate-induced colitis in mice. PLoS One. 2018;**13**(3):e0194069. DOI: 10.1371/journal.pone.0194069

[125] Sun S, Zhou S, Lei S, Zhu S, Wang K, Jiang H, et al. Jatrorrhizine reduces 5-HT and NE uptake via inhibition of uptake-2 transporters and produces antidepressant-like action in mice. Xenobiotica. 2019;**49**(10):1237-1243. DOI: 10.1080/00498254.2018.1542188

[126] Mahana D, Trent CM, Kurtz ZD, Bokulich NA, Battaglia T, Chung J, et al. Antibiotic perturbation of the murine gut microbiome enhances the adiposity, insulin resistance, and liver disease associated with high-fat diet. Genome Medicine. 2016;**8**(1):48. DOI: 10.1186/s13073-016-0297-9

[127] Liu X, Zheng H, Lu R, Huang H, Zhu H, Yin C, et al. Intervening effects of total alkaloids of *Corydalis saxicola* bunting on rats with antibiotic-induced gut microbiota dysbiosis based on 16S rRNA gene sequencing and untargeted metabolomics analyses. Frontiers in Microbiology. 2019;**10**:1151. DOI: 10.3389/fmicb.2019.01151

[128] Zamzow DR, Elias V, Legette LL, Choi J, Stevens JF, Magnusson KR. Xanthohumol improved cognitive flexibility in young mice. Behavioural Brain Research. 2014;**275**:1-10. DOI: 10.1016/j.bbr.2014.08.045

[129] Legette LL, Luna AY, Reed RL, Miranda CL, Bobe G, Proteau RR, et al. Xanthohumol lowers body weight and fasting plasma glucose in obese male Zucker fa/fa rats. Phytochemistry. 2013;**91**:236-241. DOI: 10.1016/j.phytochem.2012.04.018

[130] Hanske L, Loh G, Sczesny S, Blaut M, Braune A. Recovery and metabolism of xanthohumol in germ-free and human microbiota-associated rats. Molecular Nutrition & Food Research. 2010;**54**(10):1405-1413. DOI: 10.1002/mnfr.200900517

[131] Rajilic-Stojanovic M, de Vos WM. The first 1000 cultured species of the human gastrointestinal microbiota. FEMS Microbiology Reviews. 2014;**38**(5):996-1047. DOI: 10.1111/1574-6976.12075

[132] Paraiso IL, Plagmann LS, Yang L, Zielke R, Gombart AF, Maier CS, et al. Reductive metabolism of xanthohumol and 8-prenylnaringenin by the intestinal bacterium *Eubacterium ramulus*. Molecular Nutrition & Food Research. 2019;**63**(2):e1800923. DOI: 10.1002/mnfr.201800923

[133] Possemiers S, Bolca S, Grootaert C, Heyerick A, Decroos K, Dhooge W, et al. The prenylflavonoid isoxanthohumol from hops (*Humulus lupulus* L.) is activated into the potent phytoestrogen 8-prenylnaringenin in vitro and in the human intestine. The Journal of Nutrition. 2006;**136**(7):1862-1867. DOI: 10.1093/jn/136.7.1862

[134] Keiler AM, Zierau O, Kretzschmar G. Hop extracts and hop substances in treatment of menopausal complaints. Planta

Medica. 2013;**79**(7):576-579. DOI: 10.1055/s-0032-1328330

[135] Martino E, Della Volpe S, Terribile E, Benetti E, Sakaj M, Centamore A, et al. The long story of camptothecin: From traditional medicine to drugs. Bioorganic & Medicinal Chemistry Letters. 2017;**27**(4):701-707. DOI: 10.1016/j.bmcl.2016.12.085

[136] Ramesh M, Ahlawat P, Srinivas NR. Irinotecan and its active metabolite, SN-38: Review of bioanalytical methods and recent update from clinical pharmacology perspectives. Biomedical Chromatography. 2010;**24**(1):104-123. DOI: 10.1002/bmc.1345

[137] Nagar S, Blanchard RL. Pharmacogenetics of uridine diphosphoglucuronosyltransferase (UGT) 1A family members and its role in patient response to irinotecan. Drug Metabolism Reviews. 2006;**38**(3):393-409. DOI: 10.1080/03602530600739835

[138] Tobin PJ, Dodds HM, Clarke S, Schnitzler M, Rivory LP. The relative contributions of carboxylesterase and *β*-glucuronidase in the formation of SN-38 in human colorectal tumours. Oncology Reports. 2003;**10**(6):1977-1979. DOI: 10.3892/or.10.6.1977

[139] Flieger D, Klassert C, Hainke S, Keller R, Kleinschmidt R, Fischbach W. Phase II clinical trial for prevention of delayed diarrhea with cholestyramine/levofloxacin in the second-line treatment with irinotecan biweekly in patients with metastatic colorectal carcinoma. Oncology. 2007;**72**(1-2):10-16. DOI: 10.1159/000111083

[140] Wang X-Q, Zhang A-H, Miao J-H, Sun H, Yan G-L, Wu F-F, et al. Gut microbiota as important modulator of metabolism in health and disease. RSC Advances. 2018;**8**(74):42380-42389. DOI: 10.1039/C8RA08094A

[141] Rowland I, Gibson G, Heinken A, Scott K, Swann J, Thiele I, et al. Gut microbiota functions: Metabolism of nutrients and other food components. European Journal of Nutrition. 2018;**57**(1):1-24. DOI: 10.1007/s00394-017-1445-8

[142] Wallace BD, Wang H, Lane KT, Scott JE, Orans J, Koo JS, et al. Alleviating cancer drug toxicity by inhibiting a bacterial enzyme. Science. 2010;**330**(6005):831-835. DOI: 10.1126/science.1191175

[143] Loman BR, Jordan KR, Haynes B, Bailey MT, Pyter LM. Chemotherapy-induced neuroinflammation is associated with disrupted colonic and bacterial homeostasis in female mice. Scientific Reports. 2019;**9**(1):16490. DOI: 10.1038/s41598-019-52893-0

[144] Panebianco C, Andriulli A, Pazienza V. Pharmacomicrobiomics: Exploiting the drug-microbiota interactions in anticancer therapies. Microbiome. 2018;**6**(1):92. DOI: 10.1186/s40168-018-0483-7

[145] Subhan M, Faryal R, Macreadie I. Exploitation of *Aspergillus terreus* for the production of natural statins. Jouna l of Fungi. 2016;**2**(2):13. DOI: 10.3390/jof2020013

[146] Yoo D-H, Kim IS, Van Le TK, Jung I-H, Yoo HH, Kim D-H. Gut microbiota-mediated drug interactions between lovastatin and antibiotics. Drug Metabolism and Disposition. 2014;**42**(9):1508-1513. DOI: 10.1124/dmd.114.058354

[147] Vyas KP, Kari PH, Prakash SR, Duggan DE. Biotransformation of lovastatin. II. In vitro metabolism by rat and mouse liver microsomes and involvement of cytochrome P-450 in dehydrogenation of lovastatin. Drug Metabolism and Disposition. 1990;**18**(2):218-222. Available from: http://dmd.aspetjournals.org/content/18/2/218.long

Section 3

Search for Therapeutic Approaches

Chapter 6

Is a Fecal Microbiota Transplant Useful for Treating Inflammatory Bowel Disease?

Andra-Iulia Suceveanu, Andrada Dumitru, Marilena Musat, Claudia Voinea, Felix Voinea, Irinel Parepa, Anca Pantea Stoian, Laura Mazilu and Adrian Paul Suceveanu

Abstract

Ulcerative colitis and Crohn's disease represent the major groups of idiopathic disorders in inflammatory bowel disease (IBD). The etiology includes environmental factors, genetic factors, and immune responses. The pathogenesis is diversified; however, no guaranteed curative therapeutic regimen has been developed so far. This review contains information related to pathophysiology and current treatment options for IBD. It is known that IBD is caused by tissue-disruptive inflammatory reactions of the gut wall; that is why downregulation of the immune responses allows the healing of the damaged mucosa and allows the resetting of the physiological functions of the gut back to normal. The main treatment options are still corticosteroids, immunomodulators, antibiotics, probiotics, and a series of new agents. Their effects include modulation of cytokines, neutrophil-derived factors, adhesion molecules, and reactive oxygen/nitrogen metabolites. The monoclonal antitumor necrosis factor as infliximab recombinant anti-inflammatory cytokines or related gene therapy is also used nowadays. Still, the fecal microbiota transplantation (FMT) is considered to revolutionize the therapy in IBD, considering the abnormal inflammatory response due to the complicated relationship between microbiota and the immune system. It is imperative to mention the critical role dysbiosis may have in the pathogenesis of IBDs. This review summarizes the available literature concerning the efficacy of FMT in IBDs.

Keywords: inflammatory bowel disease, microbiota, fecal transplant

1. Introduction

The current manuscript represents an overview of literature reflecting the concern of gastroenterology physicians regarding the usefulness of fecal microbiota transplant as an appropriate and successful therapy in difficult to treat IBD patients. Inflammatory bowel disease (IBD) is a relapsing, remitting, and chronic disease that causes significant morbidity. The etiology of IBD is still unclear. The phenotype, the progression, and their development are multifactorial with environment and genetics. Nowadays, studies are confirming that the microbial influence in the pathogenesis of IBD is increasing; this fact results from an inappropriate immune response towards components of the commensal microbiota. In IBD, the diversity of luminal microbiota is reduced. *Firmicutes* (bifidobacteria, lactobacillus, and *Faecalibacterium prausnitzii*) are especially decreased. On the opposite side, the mucosal adherent bacteria are increased [1, 2].

Both ulcerative colitis and Crohn's disease share many common features like bloody stools, diarrhea, fever, and abdominal pain, but each of them also has unique features. There are many differences between the two entities, the most important being the depth of involvement in the bowel wall and their location. Crohn's disease results in transmural ulceration of any portion of the gastrointestinal tract, but it affects most often the colon and terminal ileum; at the opposite side, ulcerative colitis affects the rectum, but it may extend beyond the sigmoid and into the sigmoid or include the entire colon into the cecum [1–3].

2. Inflammatory bowel diseases

2.1 Etiology and pathophysiology

The etiology of IBD is not fully elucidated, but it seems that it often occurs in genetically susceptible individuals after an inappropriate immune response to the microbiota; so, the key to the pathogenesis of IBD is represented by the intestinal immune system. Typically, the gut epithelium prevents antigen or bacteria entry into the circulation by sealed intracellular junctions. In IBDs, because of the inflammation or the loss of the primary barrier function, these intracellular junctions are defective. The Goblet cells produce mucus as an additional protective mechanism [4].

Crohn's disease is a patchy inflammatory disease affecting any part of the digestive tract, from the mouth to the anus and perianal area. Ulcerative colitis is a disease of continuity involving the rectum and colon in a centripetal manner and associating extraintestinal involvement of the skin, bones, or eyes [5].

2.2 Histopathology

In the active form of IBD, the microscopic evaluation reveals the infiltration of lamina propria with a mix of macrophages, dendritic cells, neutrophils, and natural killer T cells. Because of the increased activation and number of these cells, there is also an increasing level of cytokines, interleukin 1b, tumor necrosis factor-alpha, and interleukin 23-TH 17 [6–8].

2.3 Clinical manifestations

Patients with ulcerative colitis describe tenesmus, abdominal pain, and a sensation of incomplete evacuation. Still, the most upsetting symptom is represented by bloody diarrhea with or without mucus. The physical exam reveals predominantly left upper or left lower quadrant abdominal pain [1].

Patients with Crohn's disease have different clinical manifestations, depending on the region of gastrointestinal involvement. Symptoms like right lower quadrant pain, non-bloody diarrhea, and weight loss are the most common in Crohn's disease [9].

2.4 Diagnosis

When diagnosing IBD, the clinician has to combine inflammatory laboratory markers, clinical findings, endoscopic biopsies, and imaging findings. The guiding laboratory tests are represented by leukocytosis, microcytic anemia, thrombocytosis, and elevated levels of C-reactive protein and erythrocyte sedimentation rate [10–12].

The imagistic methods that are useful when diagnosing IBD or assessing complications are ultrasound, magnetic resonance imaging (e.g., rectal fistulas), and computed tomography (e.g., perforation or bowel obstruction). But to confirm a diagnosis of IBD biopsies obtained via colonoscopy or esophagogastroduodenoscopy is necessarily needed [12, 13].

2.5 Options of treatment

There are many conventional and novel drug treatments that have proven efficacy in IBDs (aminosalicylates, steroids, biological therapies, and immunosuppressants). However, many patients become refractory to standard management of the disease, some of them presenting significant side effects that even require surgery. There are an increasing number of patients that live with mild active symptoms, despite medical treatment, having a poor quality of life [14–16].

The gastrointestinal microbiota has a dominant role in driving inflammation in IBD; that is why medications that manipulate the microbiota have been investigated (e.g., probiotics, prebiotics) with variable evidence of their efficacy [15, 16].

Alternative treatment in IBD management is fecal microbiota transplantation (FMT) consisting of the transfer of gut microbiota from a healthy donor, via infusion of a liquid stool suspension, to restore the intestinal microbiota of a diseased person. This therapy procedure was firstly documented in 1958 [17].

The reports concerning the patients that have been subjected to fecal microbiota transplantation had positive outcomes [18–20].

3. Microbiota

In humans, the gut microbiota varies across the digestive tract. There are relatively few bacterial species present in the stomach and small intestine, compared to the colon, which is the habitat of the highest microbial density—up to 10^{12} cells per gram of intestinal content. Ninety-nine percent of the bacteria are anaerobes, except the cecum, where aerobic bacteria achieve high frequencies [21].

Although many species in the human gut have not been studied because they cannot be cultured by the ways yet discovered, there are four dominant phyla: *Proteobacteria*, *Bacteroidetes*, *Actinobacteria*, and *Firmicutes*. Most bacteria belong to the genera *Bifidobacteria*, *Clostridium*, *Peptostreptococcus*, *Bacteroides*, *Peptococcus*, *Ruminococcus*, *Eubacterium*, and *Faecalibacterium* [22, 23].

3.1 Normal functions

The gut microbiota is still intensively studied. Besides metabolizing indigestible food compounds, it stimulates the gut immune system, directly defending against pathogens. It has a substantial role in developing and maintaining the intestinal epithelium and inducing antibody production. Also, studies focused on its action in the gut-brain axis [23, 24].

3.1.1 Direct inhibition of pathogenic bacteria

The intestinal barrier provides protection against pathogenic invasion through many defense mechanisms, including butyrate and other metabolically protective products, the commensal bacteria competitively preventing pathogenic bacteria's colonization. Disruption of intestinal barrier function may cause local or even systemic immune over-response, afferent vagus nerve activation, neuroinflammation, and mast cell degranulation. Furthermore, some species of the commensal bacteria, like *Lactobacillus plantarum*, binds and stimulates the Toll-like receptor 2 (TLR2) in the intestinal epithelium, thus maintaining epithelial integrity [24–26].

3.1.2 Development of the immune system

The human intestinal flora develops in the first 2 years after birth, when the intestinal epithelium and mucosal barrier evolve in a tolerant and even in a supportive manner. Particularly, goblet cells—the ones producing the mucosa—proliferate, resulting in a thickening of the mucosa layer in which commensal bacteria may anchor and feed but cannot penetrate. Furthermore, the gut-associated lymphoid tissue, which is part of the gut epithelium, has a role in detecting and reacting to pathogens, being tolerant of commensal species and its metabolites and digestive products of food [26–29].

Cytokines stimulate the immune system to produce inflammation to protect itself, which may decrease the immune response to control homeostasis and favor healing after injuries. There is some bacterial species in the intestinal flora which seems to drive the production of selective cytokines by the immune system, such as *Bacteroides fragilis* and *Clostridium* species, which may induce an anti-inflammatory response, although some segmented filamentous bacteria cause the production of inflammatory cytokines [29, 30].

Another function of the intestinal flora is driving the immune system to produce antibodies. Thus, B cells switch class to IgA in another way, normally needing activation from T helper cells. Intestinal epithelial cells induce NF-kB signaling, which causes the secretion of further signaling molecules. These interact with B cells and induce the switching class to IgA—an important type of antibody because it keeps healthy a mucosal environment by eliminating the microorganisms that cause inflammatory responses [24, 25].

Gut flora can produce metabolites that can affect cells in the immune system, such as short-chain fatty acids produced through fermentation, and can induce an increased production of eosinophils, neutrophils, and basophils, which are components of the innate immune system and have a role in limiting the infection's spreading [26].

3.1.3 Metabolism

The intestinal flora is essential for digestion through some enzymes that the human body does not possess to break down polysaccharides. Carbohydrates are turned into short-chain fatty acids by saccharolytic fermentation, including acetic acid, propionic acid, and butyric acid, used by the host cells as a source of energy and nutrients. Also, gut flora facilitates the absorption of minerals—magnesium, calcium, iron—and synthesizes vitamins, biotin and folate [27].

3.1.4 Microbiome-gut-brain axis

The microbiome-gut-brain axis includes the central nervous system and the neuroendocrine and neuroimmune systems, hypothalamic–pituitary–adrenal axis, sympathetic and parasympathetic arms of the autonomic nervous system, the enteric nervous system, the vagus nerve, and the intestinal microbiota [28].

The term refers to the biochemical signaling between the central nervous system and the gastrointestinal tract [29].

3.2 Dysbiosis

Dysbiosis represents microbial imbalance or maladaptation that can be caused by many triggers, such as antibiotic treatments, alcohol abuse, or inappropriate diet. Also, microorganisms present in the digestive tract may contribute to inflammatory disorders, or specific metabolites influence some signaling pathways leading to obesity and colon cancer. Additionally, sepsis may occur in cases of breaking down of the intestinal epithelium with the invasion of flora components into host's compartments [30, 31].

4. Fecal microbiota transplant (FMT)

4.1 General considerations

The new key in treating dysbiosis is the fecal microbiota transplant, which restores colonic microflora. It involves the transplantation of fecal bacteria from a healthy individual by colonoscopy, enema, orogastric tube, or orally by capsules containing freeze-dried material. FMT has been used in treating *Clostridium difficile* infection and experimentally in inflammatory bowel disease, irritable bowel syndrome, constipation, and some neurological conditions, like multiple sclerosis and Parkinson's disease [32].

The size of samples has to range from 30 to 100 grams of fecal material for the treatment to be effective. The fresh stool is needed to increase the viability of bacteria, and samples are prepared within 6–8 hours, diluted with 2.5–5 times the sample's volume with normal saline, sterile water, or 4% milk [32, 33].

The donor selection is strict and involves screening of medical history, screening for chronic diseases, and laboratory testing for pathogenic gastrointestinal infections [33].

However, clinical trials report cases of important adverse events after fecal microbiota transplant, such as gram-negative bacteremia combined with aspiration pneumonia or even toxic megacolon. Adverse events are the reason why the FDA decided to expand donor-stool screening by including tests for human T-lymphotropic virus, norovirus, and extended-spectrum beta-lactamase-producing microorganisms. Also, to minimize the risk of infection, clinicians should forget the "one size fits all" approach and consider the risks and benefits for each individual, especially in cases of immunocompromised patients [34–36].

The process of choosing donors include four stages as follows: stage 1, eliminating overweight (body mass index >30) patients, smokers, and those unable to donate periodically; stage 2, eliminating donors with microbiome-associated conditions, such as metabolic, gastrointestinal, autoimmune, allergic, atopic, neurologic, and psychiatric; stage 3, stool and nasal screening, involving tests for antibiotic-resistant bacteria; and stage 4, blood tests [35, 36].

We can conclude that healthy donors are hard to find. Thus clinicians should continue improving donor screening to reduce drug-resistant microorganisms transmission, and research should focus on the benefits and the risks involved in fecal microbiota transplant [37].

4.2 FMT in inflammatory bowel disease

It is mandatory to mention the important role that dysbiosis may have in the pathogenesis of inflammatory bowel disease, considering the abnormal inflammatory response resulted from the complex relationship between microbiota and the immune system. This feature is the reason why ongoing research carries so much interest in correcting dysbiosis using fecal microbiota transplantation [37, 38].

FMT can reduce IBD's severity by increasing the production of short-chain fatty acids (butyrate); this way, the bowel permeability is reduced, and the integrity of the gut epithelium is maintained. Also, inhibiting the production of inflammatory elements, leukocyte adhesion, and the activity of T cells, FMT may restore the immune system [38].

In patients with ulcerative colitis and Crohn's disease, preliminary clinical studies showed a long-term follow-up clinical remission maintained, even endoscopic and histologic remission in a few other cases. A meta-analysis of nine studies showed a remission rate of 36.2%. Still, the results depend on some factors like age (higher remission in younger patients with ages between 7 and 20 years old), route of administration (naso-jejunal tube, enema, colonoscopy), and dose and preparation of donor feces. Also, it seems like fecal microbiota transplant is more effective in Crohn's disease than in ulcerative colitis with remission rates of 60.5 and 22%, respectively. On the other hand, a study involving 15 patients with steroid-dependent ulcerative colitis who received fecal microbiota transplant through colonoscopy showed a long-term maintained remission in 57% of cases [38–40].

Because of the lack of uniformity regarding the treatment protocols and the delivery method, it is hard to offer a solid conclusion referring to the safety and efficacy of fecal microbiota transplant in inflammatory bowel disease. If compared with the results collected in cases of recurrent *Clostridium difficile* infection (remission in about 90%), these results may look discouraging. Still, we have to keep in mind that the inflammatory bowel disease's pathogenesis is not purely driven by dysbiosis as it happens in *Clostridium difficile* infection. Following this direction, we need more randomized controlled placebo studies to clarify the role of fecal microbiota transplant in inflammatory bowel disease [40–42].

4.3 Potential adverse effects of FMT

These can be classified into short-term and long-term side effects [19].

Short-term side effects are related to the delivery method. They may include mild fever, flatulence, constipation, diarrhea, vomiting, and abdominal discomfort, but all of these usually resolve in a few weeks. In cases when FMT was administered through the naso-jejunal tube, patients presented with high fever and the rise of the C-reactive protein. When using colonoscopy, there have been reported cases of perforation, bleeding, and symptoms related to anesthesia [43].

Due to the lack of research, there are few data collected about the dominant concern regarding the safety of fecal microbiota transplant—long-term side effects. We can speculate a considerable risk for chronic diseases, involving obesity, diabetes, colon cancer, and atherosclerosis, due to the alteration of intestinal microbiota [43, 44].

5. Conclusions

Inflammatory bowel diseases are chronic, relapsing intestinal disorders, with pathogenesis not fully elucidated. Treatment disappointments are still high, despite the availability of different therapeutic options. Patients' reduced compliance, the impoverished life of quality, and the increased economic, sanitary, and social burden worldwide are still unresolved issues. For that reason, research must continue to identify more information about the intestinal microbiota, metabolic pathways, and microbial genes. A moving target may be the identification and isolation of an active component of the microbiota that could be the new target of therapies for inflammatory bowel diseases [44–45].

Conflict of interest

All authors declare "no conflict of interest."

Author details

Andra-Iulia Suceveanu[1*], Andrada Dumitru[1], Marilena Musat[1], Claudia Voinea[1], Felix Voinea[1], Irinel Parepa[1], Anca Pantea Stoian[2], Laura Mazilu[1] and Adrian Paul Suceveanu[1]

1 Faculty of Medicine, Ovidius University, Constanta, Romania

2 University of Medicine and Pharmacy Carol Davila, Bucharest, Romania

*Address all correspondence to: andrasuceveanu@yahoo.com

References

[1] Maaser C, Sturm A, Vavricka SR, Kucharzik T, Fiorino G, Annese V, et al. ECCO-ESGAR guideline for diagnostic assessment in IBD part 1: Initial diagnosis, monitoring of known IBD, detection of complications. Journal of Crohn's and Colitis. 2019;**13**(2):144-164

[2] Sturm A, Maaser C, Calabrese E, Annese V, Fiorino G, Kucharzik T, et al. ECCO-ESGAR guideline for diagnostic assessment in IBD part 2: IBD scores and general principles and technical aspects. Journal of Crohn's & Colitis. 2019;**13**(3):273-284

[3] Trivedi PJ, Adams DH. Chemokines and chemokine receptors as therapeutic targets in inflammatory bowel disease; pitfalls and promise. Journal of Crohn's & Colitis. 2018;**12**(suppl_2):S641-S652

[4] Law CC, Sasidharan S, Rodrigues R, Nguyen DD, Sauk J, Garber J, et al. Impact of specialized inpatient IBD care on outcomes of IBD hospitalizations: A cohort study. Inflammatory Bowel Diseases. 2016 Sep;**22**(9):2149-2157

[5] Viganò CA, Beltrami MM, Bosi MF, Zanello R, Valtorta M, Maconi G. Alexithymia and psychopathology in patients suffering from inflammatory bowel disease: Arising differences and correlations to tailoring therapeutic strategies. Frontiers in Psychiatry. 2018;**9**:324

[6] Faderl M, Noti M, Corazza N, Mueller C. Keeping bugs in check: The mucus layer as a critical component in maintaining intestinal homeostasis. IUBMB Life. 2015;**67**(4):275-285

[7] Sokol H. Toward rational donor selection in faecal microbiota transplantation for IBD. Journal of Crohn's & Colitis. 2016;**10**(4):375-376

[8] Li SS, Zhu A, Benes V, et al. Durable coexistence of donor and recipient strains after fecal microbiota transplantation. Science. 2016;**352**(6285):586-589

[9] Khanna S, Tosh PK. A Clinician's primer on the role of the microbiome in human health and disease. Mayo Clinic Proceedings. 2014;**89**(1):107-114

[10] Yoon MY, Lee K, Yoon SS. Protective role of gut commensal microbes against intestinal infections. Journal of Microbiology. 2014;**52**(12):983-989

[11] Peterson LW, Artis D. Intestinal epithelial cells: Regulators of barrier function and immune homeostasis. Nature Reviews Immunology. 2014;**14**(3):141-153

[12] Honda K, Littman DR. The microbiota in adaptive immune homeostasis and disease. Nature. 2016;**535**(7610):75-84

[13] Levy M, Thaiss CA, Elinav E. Metabolites: Messengers between the microbiota and the immune system. Genes & Development. 2016;**30**(14):1589-1597

[14] Mayer EA, Knight R, Mazmanian SK, Cryan JF, Tillisch K. Gut Microbes and the Brain: Paradigm Shift in Neuroscience. The Journal of Neuroscience. 2014; **34**(46):15490-15496

[15] Kazerouni A, Burgess J, Burns LJ, Wein LM. Optimal screening and donor management in a public stool bank. Microbiome. 2015;**3**:75

[16] Cammarota G, Ianiro G, Tilg H, Rajilić-Stojanović M, Kump P, Satokari R, et al. European consensus conference on fecal microbiota transplantation in clinical practice. Gut. April 2017;**66**(4): 569-580

[17] Shen ZH, Zhu CX, Quan YS, et al. Relationship between intestinal

microbiota and ulcerative colitis: Mechanisms and clinical application of probiotics and fecal microbiota transplantation. World Journal of Gastroenterology. 2018;**24**(1):5-14

[18] Vaughn BP, Vatanen T, Allegretti JR, et al. Increased intestinal microbial diversity following fecal microbiota transplant for active Crohn's disease. Inflammatory Bowel Diseases. 2016;**22**(9):2182-2190

[19] Colman RJ. Rubin DT J fecal microbiota transplantation as therapy for inflammatory bowel disease: A systematic review and meta-analysis. Crohn's Colitis. 2014;**8**(12):1569-1581

[20] He Z, Li P, Zhu J, et al. Multiple fresh fecal microbiota transplants induces and maintains clinical remission in Crohn's disease complicated with inflammatory mass. Scientific Reports. 2017;7(1):4753

[21] Peng Z, Xiang J, He Z, et al. Colonic transendoscopic enteral tubing: A novel way of transplanting fecal microbiota. Endoscopy International Open. 2016;**4**(6):E610-E613

[22] Sunkara T, Rawla P, Ofosu A, Vinaya Gaduputi. Fecal microbiota transplant – a new frontier in inflammatory bowel disease. Journal of Inflammation Research. 2018;**11**:321-328. DOI: 10.2147/JIR.S176190

[23] Moayyedi P, Yuan Y, Baharith H, Ford AC. Fecal microbiota transplantation for *Clostridium difficile*-associated diarrhoea: A systematic review of randomised controlled trials. Medical Journal of Australia. August 2017;**207**(4):166-172

[24] Grinspan AM, Kelly CR. Fecal microbiota transplantation for ulcerative colitis: Not just yet. Gastroenterology. 2015;**149**(1):15-18

[25] Vermeire S, Joossens M, Verbeke K, et al. Donor species richness determines faecal microbiota transplantation success in inflammatory bowel disease. Journal of Crohn's & Colitis. 2016;**10**(4):387-394

[26] Baxter M, Ahmad T, Colville A, Sheridan R. Fatal aspiration pneumonia as a complication of fecal microbiota transplant. Clinical Infectious Diseases. 2015;**61**(1):136-137

[27] Moos WH, Faller DV, Harpp DN, Kanara I, Pernokas J, Powers WR, et al. Microbiota and neurological disorders: A gut feeling. BioResearch Open Access. 2016;**5**(1):137-145

[28] Alang N, Kelly CR. Weight gain after fecal microbiota transplantation. Open Forum Infectious Diseases. 2015;**2**(1):ofv004

[29] Dinan TG, Cryan JF. The impact of gut microbiota on brain and behaviour. Current Opinion in Clinical Nutrition and Metabolic Care. 2015;**18**(6):552-558

[30] Rawla P, Sunkara T, Raj JP. Role of biologics and biosimilars in inflammatory bowel disease: Current trends and future perspectives. Journal of Inflammation Research. 2018;**11**:215-226

[31] Miquel S, Leclerc M, Martin R, et al. Identification of metabolic signatures linked to anti-inflammatory effects of Faecalibacterium prausnitzii. MBio. 2015;**6**(2):pii.e00300-pii.e00315

[32] Gevers D, Kugathasan S, Denson LA, et al. The treatment-naive microbiome in new-onset Crohn's disease. Cell Host & Microbe. 2014;**15**(3):382-392

[33] Quévrain E, Maubert MA, Michon C, et al. Identification of an anti-inflammatory protein from Faecalibacterium prausnitzii, a commensal bacterium deficient in Crohn's disease. Gut. 2016;**65**(3):415-425

[34] Costello SP, Tucker EC, La Brooy J, Schoeman MN, Andrews JM. Establishing a fecal microbiota transplant service for the treatment of Clostridium difficile infection. Clinical Infectious Diseases. 2016;**62**(7):908-914

[35] Chen L, Wang W, Zhou R, et al. Characteristics of fecal and mucosa-associated microbiota in Chinese patients with inflammatory bowel disease. Medicine (Baltimore). 2014;**93**(8):e51

[36] Kelly CR, Khoruts A, Staley C, et al. Effect of fecal microbiota transplantation on recurrence in multiply recurrent Clostridium difficile infection: A randomized trial. Annals of Internal Medicine. 2016;**165**(9):609-616

[37] Moayyedi P, Surette MG, Kim PT, et al. Fecal microbiota transplantation induces remission in patients with active ulcerative colitis in a randomized controlled trial. Gastroenterology. 2015;**149**(1):102-109

[38] Rossen NG, Fuentes S, van der Spek MJ, et al. Findings from a randomized controlled trial of fecal transplantation for patients with ulcerative colitis. Gastroenterology. 2015;**149**(1):110-118

[39] Cammarota G, Ianiro G, Gasbarrini A. Fecal microbiota transplantation for the treatment of Clostridium difficile infection: A systematic review. Journal of Clinical Gastroenterology. 2014;**48**(8):693-702

[40] Youngster I, Russell GH, Pindar C, Ziv-Baran T, Sauk J, Hohmann EL. Oral, capsulized, frozen fecal microbiota transplantation for relapsing Clostridium difficile infection. Journal of the American Medical Association. 2014;**312**(17):1772-1778

[41] Li YT, Cai HF, Wang ZH, Xu J, Fang JY. Systematic review with meta-analysis: Long-term outcomes of faecal microbiota transplantation for Clostridium difficile infection. Alimentary Pharmacology & Therapeutics. 2016;**43**(4):445-457

[42] Zhang F, Cui B, He X, et al. Microbiota transplantation: Concept, methodology and strategy for its modernization. Protein & Cell. 2018;**9**(5):462-473

[43] Suskind DL, Brittnacher MJ, Wahbeh G, et al. Fecal microbial transplant effect on clinical outcomes and fecal microbiome in active Crohn's disease. Inflammatory Bowel Diseases. 2015;**21**(3):556-563

[44] Vaughn BP, Gevers D, Ting A, Korzenik JR, Robson SC, Moss AC. Fecal microbiota transplantation induces early improvement in symptoms in patients with active Crohn's disease. Gastroenterology. 2014;**146**(5 suppl 1): S591-S592

[45] Cui B, Feng Q, Wang H, et al. Fecal microbiota transplantation through mid-gut for refractory Crohn's disease: Safety, feasibility, and efficacy trial results. Journal of Gastroenterology and Hepatology. 2015;**30**(1):51-58

Chapter 7

Probiotic Bacteria in Microbiome against Allergy

Najaf Allahyari Fard, Zakie Mazhary and Nahid Javanshir

Abstract

According to the World Allergy Organization (WAO), approximately 20% of the global population suffers from allergies. As per ongoing investigations, their pervasiveness is expanding comprehensively. Allergic diseases are significant because of the high prevalence and constant increase in their costs and adverse effects on human life. Probiotics are proposed as an intervention for the prevention and treatment of allergic diseases. Various mechanisms are considered for the anti-allergic effects of probiotic properties, like detecting related molecular patterns, including DNA motifs or lipopolysaccharides (LPS) of the bacteria, through interaction with host immune systems by Toll-like receptors. In this chapter, the microbiome, allergy, and the role of immunomodulatory probiotics against allergy are discussed.

Keywords: probiotic, microbiome, allergy

1. Introduction

In industrialized countries, more than 20% of the population has symptoms of allergies. The commonness of childhood asthma increased by 50% in the USA from 1980 to 2000. The allergy mechanism is an immune response to the allergen, which is often mediated by the immunoglobulin E (IgE) antibody [1]. Allergies can be a serious risk for individuals. Allergens or pollens represent a small fraction of the proteins that humans are regularly exposed to. The importance of the topic in the uncertainty is the cause of the B and T cells' responses to these proteins [2, 3]. Notably, some proteins that are structurally similar pollens may lead to immune response, known as cross-reactivity [4].

The human body microbiome has a diverse composition of bacteria, archaea, fungi, protozoa, and viruses, which are inhabited mainly in the different epidermal surfaces of the body—the skin and mucosal surface. Some of the species of these microbiotas are identified based on cultural techniques, but due to limitations of these techniques [5], it is suggested that the number of human microbiota exceeds 1000 species or 10 times the number of cells in the entire body with 30 times larger total genome than the human genome.

A majority of these microbiotas are in the gastrointestinal tract, the major source of microbial exposure, and live in symbiosis with their host cells [6, 7]. Given up genes necessary for the survival of the commensal microbiota in other microenvironments and retained genes beneficial for the host with no or little benefit to themselves [8] are the evidences of the symbiotic coevolution of the microbiota and human [9].

The interplay of the immune system with gut microbiota starts from the day of birth and even before that. Early exposure during plasticity and prenatal period seems to be beneficial to prevent the T helper cell type-2 (Th2)-mediated allergic disease [10]. Th2 phenotype is the dominant one in newborns [11] to prevent rejection in utero. Skewing to Th2 in the immune system leads to the stimulated secretion of IgE by B cells and hence to allergies as seen in germ-free mice with the same condition that results in greater IgE responses to food antigens and failure in producing the proper amount of regulatory T cell (Treg) responses [12–14]. On the other hand, upsurge in the amount of T helper cell type-1 (Th1) also mediates the autoimmune disease [12, 15–19].

Restoring Th1/Th2 is the significant role of the microbiota [20]. The association of microbiota and the immune system is mutual. This engagement results in different signaling pathways through the immune system's molecules that increase immune responses [21]. These regulations are crucial for maintaining the homeostasis of the host and for the prevention of different diseases by inducing secretion of IgA and regulatory T cell (Treg) and stimulation of tolerance in face of common antigens [22]. So the formation, maintenance, and heterogeneity of microbiota are necessary during early life owing to their regulatory and tolerance properties in the immune system [23, 24], as it was confirmed that the lack of microflora during a short time in early life results in defection in immune regulation [15]. The mechanisms of oral tolerance which are necessary to suppress excessive immune reactions to antigens are mediated by Foxp3þ Treg [25] and IgA, which is known as the most abundant immunoglobulin and is vital in establishing the composition of microbiota [26] and strengthening the mucosal barrier function [27].

Although, it is observed that abnormal IgA responses lead to allergy development [28]. So the obligation of equilibrium of the allergy mediators is more sensible now.

Lack of genetic elements such as Toll-like receptors that cause enterocyte proliferation like TLR4 and CD14, which enhance the detection of bacterial LPS by TLR4, and TLR9, which identify the genetic molecules of the microorganisms, also increase susceptibility to allergies [29, 30].

As the priority of the microbiome is proven, some factors are mentioned as follows, to support their presence and diversity in the body. Mode of birth; surgical or natural delivery, the process of contacting microflora in the first moment of the presence. Breast or formula feeding; the extension of contact with microflora. Nutritional patterns; the habit of food, based on people's patterns to eat fatty and fast foods or healthy ones like prebiotics which are considered beneficial for even the microbiota of the host. Antibiotics; the matter of using antibiotics at an early age or the trouble of overuse of them in all ages which impair normal flora. Locality; living in urban areas with all of the stresses, less interaction with nature in contrast with living in rural areas results in losing ancient commensal microbiota. Environmental factors; contacting people or animals. Hygiene; the obsession behaviors or normal ones. Lifestyle; the matter of activity or sedentariness in someone's lifestyle.

Natural delivery and breastfeeding are the first two initial and essential exposures when the immune system is not still mature and needs antigens to active oral tolerance [18, 25, 31]. Contravention of these simple factors grounds reformed patterns of early settlement which may result in the incidence of allergy [32]. Food sensitization especially milk allergy and atopic eczema are examples of reduced gut microbial diversity [33–37].

2. The function of the intestinal microbiota

The microbiome is considered as an active organ because of manufacturing intrinsic signals for shifting postnatal development, inspiration of tolerance mechanisms and immunogenicity reduction, and resistance against invasive pathogens [38–42].

Consuming substrates of the microbiota containing fibers and mucins provides additional energy for the host as fatty acids [43]. Amines, sulfides, and ammonia are the products of them, which are detrimental metabolites for the human.

The protective barrier function against the invasive microbes by their colonization in the intestine is another potential of the microbiota. Different mechanisms for the resistance colonization of the microbiota are considered, such as competition for nutrients and connection to the binding sites and secretion of the antimicrobial substances [44].

Stimulation of the innate signaling pathways through the straight cell-to-cell communications or secretion of short-chain fatty acids (SCFA) are the other regulatory actions of the microbiota. SCFAs produced by the microbiota can direct intestinal Treg cells and inhibit pro-inflammatory responses [45–48].

3. The intestinal microbiota of allergic ones

The role of maternal microbiota in the process of preventing allergy has been proven. Infants from allergic parents are at least twice more likely to the risk of developing allergic diseases than nonallergic parents. Microbiota diversity exists between allergic and nonallergic persons. Reduction in the fecal diversity of the bacteroidetes in pregnancy is connected with the prevalence of atopic eczema in their young children [49]. The microbiota of healthy infants with nonallergic parents frequently consists of healthy lactobacilli, representing the role of maternal microbiota in preventing allergic disorders. A decrease in the number of lactobacilli and bifidobacteria and an increase in the colonization of *Staphylococcus aureus* and *Clostridium difficile* are associated with the development of allergic disorders later in life, which shows the abnormality even before the onset of the allergy [50, 51]. Apart from quantitative alterations, qualitative alterations are important in the microbiota. For example, the microbiota of infants suffering from atopic dermatitis (AD) consists of mostly *B. adolescentis* which is mainly forming adult microbiota, whereas *B. bifidum* is the fundamental former of the healthy breastfeeding infants [52, 53]. Bifidobacteria of atopic dermatitis infants encourage the secretion of pro-inflammatory cytokines, whereas the bifidobacterium of healthy ones encourages the secretion of anti-inflammatory cytokines [54]. Besides, these bifidobacteria have different adhesion behaviors to Caco-2 tissue culture cells [55] and intestinal mucus [56], which seems to be the reason for the reduction in stimulating the immune system. And at last, the metabolic activity of the microbiota composition is different too. Higher levels of butyrate, isovalerate, and caproate in the fecal matter of children with high risk for developing allergy in comparison with normal children are the confirmation of this claim [57].

4. Stabilizing intestinal bacterial flora

The intestine, the largest immune organ of the body, which is the source of the most antibody-producing cells [58] is the target of triggering maturation of the immune system or the restoration of the impaired commensal bacteria. Stimulation of the immune system is one of the most impressive functions of the resident microbiota of the intestine. Probiotic bacteria are considered as a safe solution for modulation of diminished commensal composition and also as influencer of the immune system in preventing allergic disorders [59]. Lactic acid bacteria and bifidobacteria are good candidates as probiotics with an appropriate life span, no toxic or pathogenic properties, and no inflammatory induction. The selection of the bacteria as probiotics is mainly based on no harmful side effects during the history of their use for a long time. Consumption of these probiotics aid in balancing the ratio of the intestinal flora, avoiding the inhabiting of the pathogens by preventing the binding of them to the host cells, and suppressing the inflammation, which all are as the result of immune system regulation [60]. The effects of probiotics vary with the dose, strain, and duration of consumption and timing.

But the problem of the probiotics is their longevity and residence in the body of the host, as it was seen that they only remain during the administration period and not after that, showing the transient colonization of the probiotics [61–65].

Long-term effects of probiotics in different periods of everyone's life need to be more investigated in complementary studies.

As it was mentioned, immune tolerance is one of the necessary immune reactions to stop excessive inflammatory reactions. Preservation of this tolerance involves the integrity of the epithelial barrier that is heightened by commensal anaerobes, such as *Clostridium* spp. [15, 16, 66, 67], *Bacteroides fragilis* [68], and *Clostridium* spp. [66, 67], which are potent inducers, persuading Foxp3þ Treg differentiation to maintain mucosal tolerance and intestine integrity. Clostridia class also has adaptation properties in the intestinal cells to the routine exposure of an extended range of the antigens. This adaptation is acquired by the effect of IL-22 secreted by innate lymphoid cells which control enterocyte proliferation, activating the secretion of the mucus and antimicrobial production [16, 69].

5. Probiotics' mechanisms of action in allergic disorders

Immune homeostasis develops in the gut as a relationship between the intestinal microbiota, the luminal antigens, and the epithelial barrier is established. Microbial intestinal colonization starts after conception. This happens when the newborn's sterile gut is slowly colonized by environmental bacteria and by interaction with the mother's intestinal flora and surroundings and probably by genetic factors [70–72]. Exposure to microbial flora early in life causes a transition in the T helper cell type-1 (Th1)/Th2 cytokine balance, promoting a Th1 cell response [73].

An infant's immune system at birth is not completely formed and appears to be geared toward a Th2 phenotype to prevent in utero rejection [74]. Nevertheless, the Th2 phenotype results in a stimulated production of IgE by B cells and therefore raises the risk of allergic reactions by mast cells activation [75, 76]. Early in life microbial stimulation will reverse the Th2 bias and promote the expansion of the Th1 phenotype and promote Th3 cell activity [76]. In this way, their combined activity will lead to B-cells releasing IgA. IgA contributes to the elimination of allergens and hence would reduce the immune system's response to antigens. Th1 phenotype-produced cytokines will also reduce inflammation and promote tolerance toward specific antigens [77].

The hygiene concept states that inadequate or aberrant exposure to environmental microbes is one of the triggers of allergy production and related diseases [78]. As mentioned before, allergic diseases are associated with a change in the Th1/Th2 cytokine balance leading to Th2 cytokine activation and interleukin-4 (IL-4), IL-5, and IL-13 activation as well as IgE production [79, 80]. Probiotics significantly alter the gut microenvironment by encouraging a shift in local microflora and cytokine secretion [81] and can potentially modulate enterocyte Toll-like receptors and proteoglycan recognition proteins, resulting in dendritic cell (DC) activation and a Th1 response. The resulting stimulation of Th1 cytokines can suppress reactions to Th2 [82].

6. Probiotics in atopic dermatitis

Atopic dermatitis (AD) is a widespread chronic inflammatory skin condition with a prevalence of around 20% in children and 2–5% in adults worldwide [83]. In recent years, the function of the intestinal microbiota in the aetiopathogenesis of AD has become increasingly important. Atopic dermatitis probiotic therapy is widely studied, with contradictory outcomes [84]. Probiotics containing *Lactobacillus* spp. for the treatment of infantile atopic dermatitis showed beneficial effects in children. Caution should however be raised when treating children under the age of 1 years of age [85]. In addition, mild subjects are exceptions to that beneficial effect. More studies could be informative in investigating the efficacy of *Bifidobacterium* strains. Further larger studies in the treatment of pediatric AD are also required to examine the health, dose-response profile, and long-term impact of probiotics [86].

7. Probiotics in asthma

Asthma, a chronic complex airway disease, is characterized by reversible airflow obstruction, bronchial hyper responsiveness, and underlying inflammation [87]. In recent decades, the prevalence of asthma has risen. One possible mechanism behind this high prevalence is the microbial hypothesis, which suggests that less microbial exposure upregulates T helper cell type-2 (Th2) cytokine development, leading to a rise in allergic diseases [75]. A meta-analysis found that while perinatal and early-life probiotic administration reduces children's risk of atopic sensitization and total rates of immunoglobulin E (IgE), it may not reduce their risk of asthma [88]. However, in addition to routine treatment, several studies have documented the advantage of using probiotics for treating children with asthma. A randomized, placebo-controlled trial for 7-week treatment with *Enterococcus faecalis* showed reduced peak flow variability in children with asthma [89]. Lee et al. have reported substantial improvements in the pulmonary function of children with asthma following a regimen of supplementation of vegetables, fish oil, and fruit along with probiotic administration. Studies, however, have shown that *Lactobacillus* is safe for children with asthma [90, 91].

8. Probiotics in allergic rhinitis

On these bases, probiotic bacteria are capable of altering immune responses through a range of mechanisms that could minimize allergic reactions to airborne allergens without the side effects of any current drugs, and these possible mechanisms, as shown in **Figure 1**, include regulatory T cells that dampen immune responses and suppress the production of IgE antibodies [92, 93]. There are contradictory studies about the effectiveness of probiotics in treating allergic rhinitis [94]. It is reported that *L. casei* decreased the number of episodes of rhinitis in 64 preschoolers with allergic rhinitis [95]. Nonetheless, another study found that patients treated with *Lactobacillus* GG during the birch pollen season who were allergic to birch pollen and apple food found no improvement in symptom score and no reduced sensitivity to birch pollen and apple following probiotic supplementation [96]. Probiotic consumption increased life performance in allergic rhinitis patients. Blood or immunological parameters did not alter significantly in the probiotic community. This indicates probiotics may be useful in allergic rhinitis, but the data present are not sufficient to make any guidelines for treatment [97, 98].

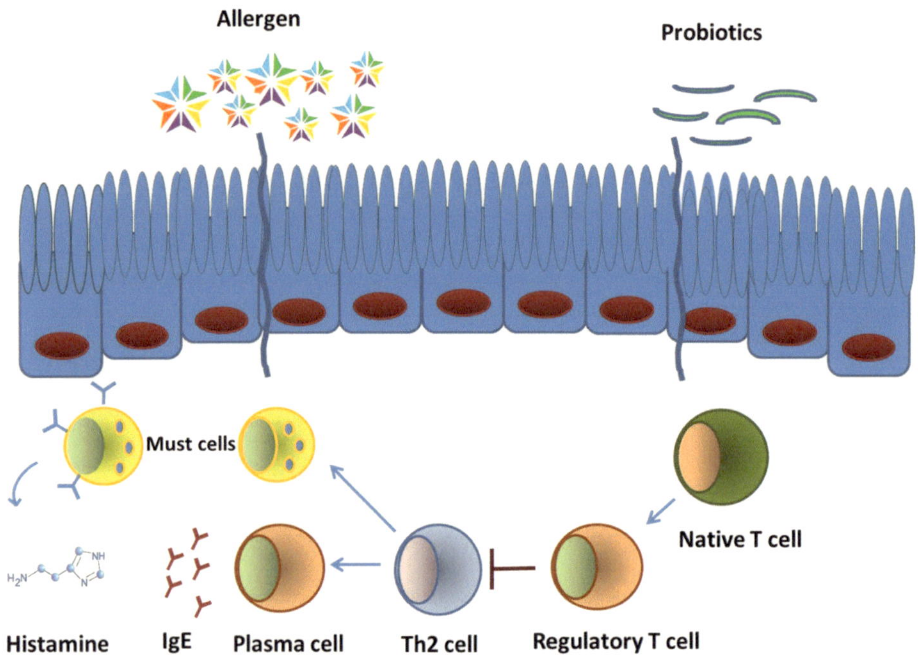

Figure 1.
Probiotic mechanism against the allergen include increasing regulatory T cells that damp down immune responses and suppress the production of IgE.

9. Probiotics in food allergy

Food allergy (FA) is one of the pediatric age's most common allergic disorders and has been considered a global health issue, particularly in the developed world.

Naturally, many subjects with FA outgrow this over time. Cow's milk allergy (CMA), hen's egg allergy, and wheat allergy resolve by 5–10 years in 50% of children. Many FAs (including peanuts, tree nuts, and fish) have low-resolution levels and are seen as persistent [99]. Furthermore, certain types of FA may be correlated with the subsequent development of other allergic symptoms such as oculorhinitis, atopic dermatitis, asthma, and urticaria (the so-called "Atopic March") [100] as well as other diseases such as functional gastrointestinal disorders (FGIDs), inflammatory intestinal diseases (IBD) [101], and psychiatric disorders such as attentive autistic spectrum disorders (ASD).

The gut microbiome-immune system axis that influences the frequency of FA may be modulated by several genetic, environmental, and dietary factors [102]. For example, increased family size, pet and/or rural exposure, balanced diet (full of fibers, fermented foods, antioxidants, omega-3), breastfeeding, and probiotic use are correlated with FA safety. Conversely, C-section, prenatal, and early-life exposure to antibiotics/gastric acid inhibitors/antiseptic agents, unhealthy diet (low fibers/high saturated fats, and junk foods) may increase the risk of developing FA. All these environmental factors mainly operate on a modulation of the structure and function of the gut microbiota, which may in effect be responsible for the epigenetic control of genes involved in immune tolerance.

The pathogenesis of these incidents also is largely unknown, but increasing evidence suggests the hypothesis that disturbance of intestinal microbiome, leading to alterations in the immune system and gut-brain axis, may affect the occurrence of FA and FA-related conditions later in life [103] (**Figure 2**).

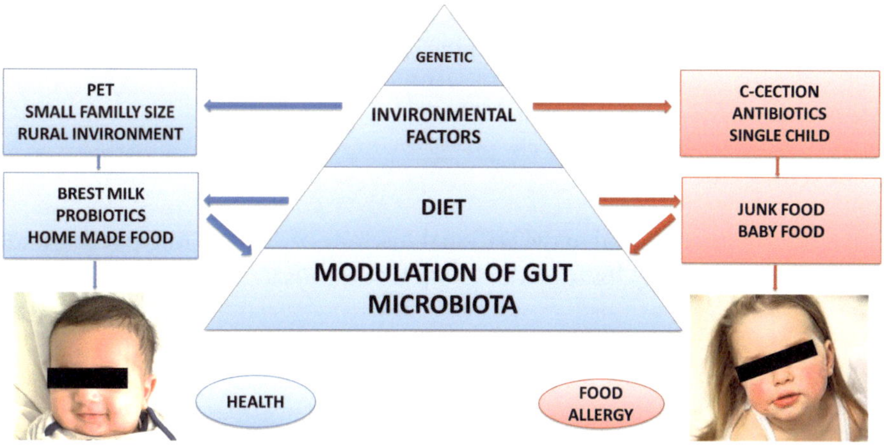

Figure 2.
Good microbiome as a target for food allergy intervention.

Mediterranean diet (MD) is described as a healthy, balanced diet. It is associated with a high intake of assorted cereals, legumes, fruit, vegetables, olive oil, and nuts; moderate consumption of red wine, poultry, and fish; and a lower intake of red meat and sweets. MD has been shown to have a protective role against allergic illnesses in children during pregnancy and early life [104].

Elevated adherence to MD was associated with increased levels of Prevotella and other Firmicutes and production of short-chain fatty acids (SCFAs) [102]. One of the strongest links between diet, gut microbiome, and allergic diseases is the immunomodulatory mechanisms elicited by SCFAs [105]. Common SCFAs contain acetate, butyrate, propionate, and valerate. SCFA-producing bacteria include *Faecalibacterium prausnitzii* and *Eubacterium rectale*. Roseburia is an important butyrate producer [106]. SCFAs are a major source of energy for colonocytes and epigenetically influence many nonimmune functions (tightened junction proteins, and mucus production) and immune functions (macrophages, neutrophils, dendritic cells (DCs), and T and B cells) [107, 108]. Enterocyte interaction of SCFAs is mediated by G-protein-coupled receptors, namely GPCRs, GPR41, GPR43, GPR109A, and Olfr78 [109, 110]. The hopeful target of novel therapeutic and preventive approaches against FA may be the gut microbiome. The results of the studies are promising, but more research is needed for the better definition of the potential for diet-gut microbiome—immune system axis modulation to counter FA. We are entering a new age in which the production and function of the immune system can be controlled by dietary intervention, and the clinical effect can be assessed by gut microbes and their metabolites. Given the current gaps in research methods and data analysis and interpretation, more scientific evidence is required which can be converted into clinical evidence praxis [103].

10. Conclusion

The results of many studies have demonstrated that there is a strong relationship between modifications within the microbiome and many diseases. Much evidence proves that healthy microbiota affects and improves the immune system. It seems that probiotics can have an important role in the prevention of many diseases such as allergy. Microbiota diversity exists between allergic and nonallergic persons. Different mechanisms are considered for the anti-allergic impact of probiotics, like detecting related molecular patterns, including DNA motifs or LPS of the bacteria by Toll-like receptors. Probiotic mechanism against the allergen includes increasing regulatory T cells that damp down immune responses and suppress the production of IgE.

Acknowledgements

The authors are thankful to the National Institute of Genetic Engineering and Biotechnology (NIGEB), Iran, for providing the facilities.

Conflict of interest

The authors declare no conflict of interest.

Author details

Najaf Allahyari Fard[1]*, Zakie Mazhary[2] and Nahid Javanshir[1]

1 National Institute of Genetic Engineering and Biotechnology, Tehran, Iran

2 Islamic Azad University, Science and Research Branch, Tehran, Iran

*Address all correspondence to: allahyar@nigeb.ac.ir

References

[1] Zhang L et al. SORTALLER: Predicting allergens using substantially optimized algorithm on allergen family featured peptides. Bioinformatics. 2012;**28**(16):2178-2179

[2] Goodman RE. Biosafety: Evaluation and regulation of genetically modified (GM) crops in the United States. Journal of Hauzhong Agricultural University. 2014;**33**(6):85-114

[3] Goodman RE et al. Allergenicity assessment of genetically modified crops—What makes sense? Nature Biotechnology. 2008;**26**(1):73-81

[4] Hayes M et al. In silico tools for exploring potential human allergy to proteins. Drug Discovery Today: Disease Models. 2015;**17**:3-11

[5] Zoetendal EG, Akkermans AD, De Vos WM. Temperature gradient gel electrophoresis analysis of 16S rRNA from human fecal samples reveals stable and host-specific communities of active bacteria. Applied and Environmental Microbiology. 1998;**64**(10):3854-3859

[6] Neish AS. Microbes in gastrointestinal health and disease. Gastroenterology. 2009;**136**(1):65-80

[7] Morelli L. Postnatal development of intestinal microflora as influenced by infant nutrition. The Journal of Nutrition. 2008;**138**(9):1791S-1795S

[8] McCutcheon JP, Moran NA. Extreme genome reduction in symbiotic bacteria. Nature Reviews Microbiology. 2012;**10**(1):13-26

[9] Stein MM et al. Innate immunity and asthma risk in Amish and Hutterite farm children. New England Journal of Medicine. 2016;**375**(5):411-421

[10] West CE et al. Probiotics for treatment and primary prevention of allergic diseases and asthma: Looking back and moving forward. Expert Review of Clinical Immunology. 2016;**12**(6):625-639

[11] Zeiger RS. Food allergen avoidance in the prevention of food allergy in infants and children. Pediatrics. 2003;**111**(Supplement 3):1662-1671

[12] West CE, Jenmalm M, Prescott S. The gut microbiota and its role in the development of allergic disease: A wider perspective. Clinical & Experimental Allergy. 2015;**45**(1):43-53

[13] Wesemann DR et al. Microbial colonization influences early B-lineage development in the gut lamina propria. Nature. 2013;**501**(7465):112-115

[14] Rodriguez B et al. Infant gut microbiota is protective against cow's milk allergy in mice despite immature ileal T-cell response. FEMS Microbiology Ecology. 2012;**79**(1):192-202

[15] Wesemann DR, Nagler CR. The microbiome, timing, and barrier function in the context of allergic disease. Immunity. 2016;**44**(4):728-738

[16] Stefka AT et al. Commensal bacteria protect against food allergen sensitization. Proceedings of the National Academy of Sciences. 2014;**111**(36):13145-13150

[17] Prescott SL, Björkstén B. Probiotics for the prevention or treatment of allergic diseases. Journal of Allergy and Clinical Immunology. 2007;**120**(2):255-262

[18] Blázquez AB, Berin MC. Microbiome and food allergy. Translational Research. 2017;**179**:199-203

[19] Björkstén B. Effects of intestinal microflora and the environment on the development of asthma and allergy.

In: Springer Seminars in Immunopathology. Springer; 2004

[20] Walker WA, Iyengar RS. Breast milk, microbiota, and intestinal immune homeostasis. Pediatric Research. 2015;**77**(1-2):220-228

[21] Nicholson JK et al. Host-gut microbiota metabolic interactions. Science. 2012;**336**(6086):1262-1267

[22] Kirjavainen PV, Gibson GR. Healthy gut microflora and allergy: Factors influencing development of the microbiota. Annals of Medicine. 1999;**31**(4):288-292

[23] Nagler-Anderson C et al. Mucosal antigen presentation and the control of tolerance and immunity. Trends in Immunology. 2001;**22**(3):120-122

[24] Nagler-Anderson C. Man the barrier! Strategic defences in the intestinal mucosa. Nature Reviews Immunology. 2001;**1**(1):59-67

[25] Yu W, Freeland DMH, Nadeau KC. Food allergy: Immune mechanisms, diagnosis and immunotherapy. Nature Reviews Immunology. 2016;**16**(12):751

[26] Pabst O, Cerovic V, Hornef M. Secretory IgA in the coordination of establishment and maintenance of the microbiota. Trends in Immunology. 2016;**37**(5):287-296

[27] Pabst O. New concepts in the generation and functions of IgA. Nature Reviews Immunology. 2012;**12**(12):821-832

[28] Dzidic M et al. Aberrant IgA responses to the gut microbiota during infancy precede asthma and allergy development. Journal of Allergy and Clinical Immunology. 2017;**139**(3): 1017-1025. e14

[29] Round JL, Mazmanian SK. Inducible Foxp3+ regulatory T-cell development by a commensal bacterium of the intestinal microbiota. Proceedings of the National Academy of Sciences. 2010;**107**(27):12204-12209

[30] Bashir MEH et al. Toll-like receptor 4 signaling by intestinal microbes influences susceptibility to food allergy. The Journal of Immunology. 2004;**172**(11):6978-6987

[31] Rodríguez JM et al. The composition of the gut microbiota throughout life, with an emphasis on early life. Microbial Ecology in Health and Disease. 2015;**26**(1):26050

[32] West CE et al. The gut microbiota and inflammatory noncommunicable diseases: Associations and potentials for gut microbiota therapies. Journal of Allergy and Clinical Immunology. 2015;**135**(1):3-13

[33] Wang M et al. Reduced diversity in the early fecal microbiota of infants with atopic eczema. Journal of Allergy and Clinical Immunology. 2008;**121**(1):129-134

[34] Ismail IH et al. Reduced gut microbial diversity in early life is associated with later development of eczema but not atopy in high-risk infants. Pediatric Allergy and Immunology. 2012;**23**(7):674-681

[35] Bunyavanich S et al. Early-life gut microbiome composition and milk allergy resolution. Journal of Allergy and Clinical Immunology. 2016;**138**(4):1122-1130

[36] Azad MB et al. Infant gut microbiota and food sensitization: Associations in the first year of life. Clinical & Experimental Allergy. 2015;**45**(3):632-643

[37] Abrahamsson TR et al. Low diversity of the gut microbiota in infants with atopic eczema. Journal of Allergy and Clinical Immunology. 2012;**129**(2):434-440. e2

[38] Olszak T et al. Microbial exposure during early life has persistent effects on natural killer T cell function. Science. 2012;**336**(6080):489-493

[39] Iweala OI, Nagler CR. Immune privilege in the gut: The establishment and maintenance of non-responsiveness to dietary antigens and commensal flora. Immunological Reviews. 2006;**213**(1):82-100

[40] Iannitti T, Palmieri B. Therapeutical use of probiotic formulations in clinical practice. Clinical Nutrition. 2010;**29**(6):701-725

[41] Hooper LV, Gordon JI. Commensal host-bacterial relationships in the gut. Science. 2001;**292**(5519):1115-1118

[42] Brandtzaeg P. Development of the mucosal immune system in humans. In: Recent developments in infant nutrition. Springer-Verlag GmbH Germany, part of Springer Nature; 1996. pp. 349-376

[43] Ouwehand AC et al. Prebiotics and other microbial substrates for gut functionality. Current Opinion in Biotechnology. 2005;**16**(2):212-217

[44] Adlerberth MC, Poilane I, Wold A, Collignon A, Ingegerd. Mechanisms of colonisation and colonisation resistance of the digestive tract part 1: Bacteria/host interactions. Microbial Ecology in Health and Disease. 2000;**12**(2):223-239

[45] Smith PM et al. The microbial metabolites, short-chain fatty acids, regulate colonic Treg cell homeostasis. Science. 2013;**341**(6145):569-573

[46] Furusawa Y et al. Commensal microbe-derived butyrate induces the differentiation of colonic regulatory T cells. Nature. 2013;**504**(7480):446-450

[47] Chang PV et al. The microbial metabolite butyrate regulates intestinal macrophage function via histone deacetylase inhibition. Proceedings of the National Academy of Sciences. 2014;**111**(6):2247-2252

[48] Arpaia N, Rudensky AY. Microbial metabolites control gut inflammatory responses. Proceedings of the National Academy of Sciences. 2014;**111**(6):2058-2059

[49] West CE et al. Gut microbiome and innate immune response patterns in Ig E-associated eczema. Clinical & Experimental Allergy. 2015;**45**(9):1419-1429

[50] Sepp E et al. Intestinal microbiota and immunoglobulin E responses in 5-year-old estonian children. Clinical & Experimental Allergy. 2005;**35**(9):1141-1146

[51] Kalliomäki M et al. Distinct patterns of neonatal gut microflora in infants in whom atopy was and was not developing. Journal of Allergy and Clinical Immunology. 2001;**107**(1):129-134

[52] Ouwehand AC et al. Differences in Bifidobacterium flora composition in allergic and healthy infants. Journal of Allergy and Clinical Immunology. 2001;**108**(1):144-145

[53] Murray C et al. Fecal microbiota in sensitized wheezy and non-sensitized non-wheezy children: A nested case–control study. Clinical & Experimental Allergy. 2005;**35**(6):741-745

[54] He F et al. Intestinal *Bifidobacterium* species induce varying cytokine production. Journal of Allergy and Clinical Immunology. 2002;**109**(6):1035-1036

[55] Morita H et al. Adhesion of lactic acid bacteria to Caco-2 cells and their effect on cytokine secretion. Microbiology and Immunology. 2002;**46**(4):293-297

[56] He F et al. Comparison of mucosal adhesion and species identification of bifidobacteria isolated from healthy and allergic infants. FEMS Immunology and Medical Microbiology. 2001;**30**(1):43-47

[57] Norin E, Midtvedt T, Björkstén B. Development of faecal short-chain fatty acid pattern during the first year of life in Estonian and Swedish infants. Microbial Ecology in Health and Disease. 2004;**16**(1):8-12

[58] Brandtzaeg P. Current understanding of gastrointestinal immunoregulation and its relation to food allergy. Annals of the New York Academy of Sciences. 2002;**964**(1):13-45

[59] Saavedra J, Tschernia A. Human studies with probiotics and prebiotics: Clinical implications. British Journal of Nutrition. 2002;**87**(S2):S241-S246

[60] Salinas I et al. Monospecies and multispecies probiotic formulations produce different systemic and local immunostimulatory effects in the gilthead seabream (*Sparus aurata* L.). Fish & Shellfish Immunology. 2008;**25**(1-2):114-123

[61] Wickens K et al. A differential effect of 2 probiotics in the prevention of eczema and atopy: A double-blind, randomized, placebo-controlled trial. Journal of Allergy and Clinical Immunology. 2008;**122**(4):788-794

[62] West CE, Hammarström ML, Hernell O. Probiotics in primary prevention of allergic disease–follow-up at 8-9 years of age. Allergy. 2013;**68**(8):1015-1020

[63] West CE et al. Effects of feeding probiotics during weaning on infections and antibody responses to diphtheria, tetanus and Hib vaccines. Pediatric Allergy and Immunology. 2008;**19**(1):53-60

[64] Rutten N et al. Long term development of gut microbiota composition in atopic children: impact of probiotics. PLoS One. 2015;**10**(9):e0137681

[65] Dotterud CK et al. Does maternal perinatal probiotic supplementation alter the intestinal microbiota of mother and child? Journal of Pediatric Gastroenterology and Nutrition. 2015;**61**(2):200-207

[66] Atarashi K et al. Induction of colonic regulatory T cells by indigenous *Clostridium* species. Science. 2011;**331**(6015):337-341

[67] Atarashi K et al. T reg induction by a rationally selected mixture of Clostridia strains from the human microbiota. Nature. 2013;**500**(7461):232-236

[68] Round JL, Mazmanian SK. The gut microbiota shapes intestinal immune responses during health and disease. Nature Reviews Immunology. 2009;**9**(5):313-323

[69] Sabat R, Ouyang W, Wolk K. Therapeutic opportunities of the IL-22–IL-22R1 system. Nature Reviews Drug Discovery. 2014;**13**(1):21-38

[70] Strachan DP. Family size, infection and atopy: The first decade of the'hygiene hypothesis'. Thorax. 2000;**55**(Suppl 1):S2

[71] Noverr MC, Huffnagle GB. The 'microflora hypothesis' of allergic diseases. Clinical & Experimental. Allergy. 2005;**35**(12):1511-1520

[72] Tang R-B. Risk factors associated with the development of asthma. Journal of the Chinese Medical Association. 2005;**68**(5):199-201

[73] Winkler P et al. Molecular and cellular basis of microflora-host interactions. Journal of Nutrition. 2007;**137**(3):756S-772S

[74] Simon AK, Hollander GA, McMichael A. Evolution of the immune system in humans from infancy to old age. Proceedings of the Royal Society B: Biological Sciences. 2015;**282**(1821):20143085

[75] Ouwehand AC. Antiallergic effects of probiotics. Journal of Nutrition. 2007;**137**(3):794S-797S

[76] Tang R-B, Chang J-K, Chen H-L. Can probiotics be used to treat allergic diseases? Journal of the Chinese Medical Association. 2015;**78**(3):154-157

[77] Paibomesai MA. Epigenetic influences on bovine T-helper 1 and T-helper 2 cytokines (interferon-gamma and Interleukin-4) in high and low immune responders around the peripartum period [Diss.]; 2017

[78] Daley D. The evolution of the hygiene hypothesis: The role of early-life exposures to viruses and microbes and their relationship to asthma and allergic diseases. Current Opinion in Allergy and Clinical Immunology. 2014;**14**(5):390-396

[79] Deo SS et al. Role played by Th2 type cytokines in IgE mediated allergy and asthma. Lung India: Official Organ of Indian Chest Society. 2010;**27**(2):66

[80] Ngoc LP et al. Cytokines, allergy, and asthma. Current Opinion in Allergy and Clinical Immunology. 2005;**5**(2):161-166

[81] Jae LB, Bak Y-T. Irritable bowel syndrome, gut microbiota and probiotics. Neurogastroenterology & Motility. 2011;**17**(3):252

[82] Rabia A, Shah NP. Immune system stimulation by probiotic microorganisms. Critical Reviews in Food Science and Nutrition. 2014;**54**(7):938-956

[83] Stephanie W et al. Effects of probiotics on atopic dermatitis: A randomised controlled trial. Archives of Disease in Childhood. 2005;**90**(9):892-897

[84] Perceval C et al. Prophylactic use of probiotics for gastrointestinal disorders in children. The Lancet Child & Adolescent Health. 2019;**3**(9):655-662

[85] Sofia R et al. Gut microbiota profile in children affected by atopic dermatitis and evaluation of intestinal persistence of a probiotic mixture. Scientific Reports. 2019;**9**(1)

[86] Fang Z et al. Strain-specific ameliorating effect of *Bifidobacterium longum* on atopic dermatitis in mice. Journal of Functional Foods. 2019;**60**:103426

[87] Hamdan AL-J et al. Asthma control and predictive factors among adults in Saudi Arabia: Results from the epidemiological study on the Management of Asthma in asthmatic Middle East adult population study. Annals of Thoracic Medicine. 2019;**14**(2):148

[88] Ahanchian H et al. Probiotics for the treatment of asthma: A systematic review and meta-analysis of randomized trials. International Journal of Pediatrics. 2020;**8**(5):11271-11285

[89] Johnson C. Asthma flare-ups in children returning to school: New blood screening device developed: More bowel cancer cases for under 50s. Australian Medicine. 2019;**31**(2):25

[90] Wei X et al. Association between probiotic supplementation and asthma incidence in infants: A meta-analysis of randomized controlled trials. The Journal of Asthma. 2020;**57**(2):167-178

[91] Wu C-T et al. Effect of *Lactobacillus rhamnosus* GG immunopathologic changes in chronic mouse asthma model. Journal of Microbiology,

Immunology and Infection. 2019;**52**(6):911-919

[92] Schaefer M, Enck P. Effects of a probiotic treatment (*Enterococcus faecalis*) and open-label placebo on symptoms of allergic rhinitis: Study protocol for a randomised controlled trial. BMJ Open. 2019;**9**:e031339. DOI: 10.1136/bmjopen-2019-031339

[93] Dehnavi S et al. A significant decrease in the gene expression of interleukin-17 following the administration of synbiotic in patients with allergic rhinitis who underwent immunotherapy: A placebo-controlled clinical trial. Journal of Research in Medical Sciences: The Official Journal of Isfahan University of Medical Sciences. 2019;**24**:1-7

[94] Sadeghi-Shabestari M et al. Effect of probiotics on allergic rhinitis: A randomized, controlled, clinical trial. Galen Medical Journal. 2020;**9**:1918

[95] Giovannini M, Agostoni C, Riva E, Salvini F, Ruscitto A, Zuccotti GV, et al. A randomized prospective double blind controlled trial on effects of long-term consumption of fermented milk containing Lactobacillus caseiin pre-school children with allergic asthma and/or rhinitis. Pediatric Research. 2007;**62**:215-220

[96] Helin T, Haahtela S, Haahtela T. No effect of oral treatment with an intestinal bacterial strain, *Lactobacillus rhamnosus* (ATCC 53103), on birch-pollen allergy: A placebo-controlled double-blind study. Allergy. 2002;**57**:243-246

[97] Yang G, Liu ZG, Yang PC. Treatment of allergic rhinitis with probiotics; an alternative approach. North American Journal of Medical Sciences. 2013;**5**:465-468

[98] Das RR, Singh M, Shafiq N. Probiotics in treatment of allergic rhinitis. World Allergy Organization Journal. 2010;**3**:239-244

[99] Marcello G et al. A randomized prospective double blind controlled trial on effects of long-term consumption of fermented milk containing *Lactobacillus casei* in pre-school children with allergic asthma and/or rhinitis. Pediatric Research. 2007;**62**(2):215-220

[100] Miguel S, Lu P, Bonilla S. Cow's-milk allergy is a risk factor for the development of FGIDs in children. Journal of Pediatric Gastroenterology and Nutrition. 2011;**52**(2):166-169

[101] Virta LJ, Kautiainen H, Kolho K-L. Symptoms suggestive of cow's milk allergy in infancy and pediatric inflammatory bowel disease. Pediatric Allergy and Immunology. 2016;**27**(4):361-367

[102] Filippis D, Francesca, et al. High-level adherence to a Mediterranean diet beneficially impacts the gut microbiota and associated metabolome. Gut. 2016;**65**(11):1812-1821

[103] Roberto BC et al. Gut microbiome as target for innovative strategies against food allergy. Frontiers in Immunology. 2019;**10**:191

[104] Castro-Rodriguez JA, Garcia-Marcos L. What are the effects of a Mediterranean diet on allergies and asthma in children? Frontiers in Pediatrics. 2017;**5**:72

[105] Craig MK et al. The nutrition-gut microbiome-physiology axis and allergic diseases. Immunological Reviews. 2017;**278**(1):277-295

[106] Louis P, Flint HJ. Diversity, metabolism and microbial ecology of butyrate-producing bacteria from the human large intestine. FEMS Microbiology Letters. 2009;**294**(1):1-8

[107] Schauber J et al. Expression of the cathelicidin LL-37 is modulated by

short chain fatty acids in colonocytes: Relevance of signalling pathways. Gut. 2003;**52**(5):735-741

[108] Luying P et al. Butyrate enhances the intestinal barrier by facilitating tight junction assembly via activation of AMP-activated protein kinase in Caco-2 cell monolayers. Journal of Nutrition. 2009;**139**(9):1619-1625

[109] Brown AJ et al. The orphan G protein-coupled receptors GPR41 and GPR43 are activated by propionate and other short chain carboxylic acids. The Journal of Biological Chemistry. 2003;**278**(13):11312-11319

[110] Nilsson NE et al. Identification of a free fatty acid receptor, FFA2R, expressed on leukocytes and activated by short-chain fatty acids. Biochemical and Biophysical Research Communications. 2003;**303**(4):1047-1052